INTERNATIONAL ENERGY AGENCY

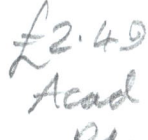

£2·49
Acad

D1807237

TACKLING INVESTMENT CHALLENGES IN POWER GENERATION

In IEA countries

INTERNATIONAL ENERGY AGENCY

The International Energy Agency (IEA) is an autonomous body which was established in November 1974 within the framework of the Organisation for Economic Co-operation and Development (OECD) to implement an international energy programme.

It carries out a comprehensive programme of energy co-operation among twenty-six of the OECD thirty member countries. The basic aims of the IEA are:

- To maintain and improve systems for coping with oil supply disruptions.
- To promote rational energy policies in a global context through co-operative relations with non-member countries, industry and international organisations.
- To operate a permanent information system on the international oil market.
- To improve the world's energy supply and demand structure by developing alternative energy sources and increasing the efficiency of energy use.
- To assist in the integration of environmental and energy policies.

The IEA member countries are: Australia, Austria, Belgium, Canada, the Czech Republic, Denmark, Finland, France, Germany, Greece, Hungary, Ireland, Italy, Japan, the Republic of Korea, Luxembourg, the Netherlands, New Zealand, Norway, Portugal, Spain, Sweden, Switzerland, Turkey, the United Kingdom and the United States. The European Commission takes part in the work of the IEA.

ORGANISATION FOR ECONOMIC CO-OPERATION AND DEVELOPMENT

The OECD is a unique forum where the governments of thirty democracies work together to address the economic, social and environmental challenges of globalisation. The OECD is also at the forefront of efforts to understand and to help governments respond to new developments and concerns, such as corporate governance, the information economy and the challenges of an ageing population. The Organisation provides a setting where governments can compare policy experiences, seek answers to common problems, identify good practice and work to co-ordinate domestic and international policies.

The OECD member countries are: Australia, Austria, Belgium, Canada, the Czech Republic, Denmark, Finland, France, Germany, Greece, Hungary, Iceland, Ireland, Italy, Japan, Korea, Luxembourg, Mexico, the Netherlands, New Zealand, Norway, Poland, Portugal, the Slovak Republic, Spain, Sweden, Switzerland, Turkey, the United Kingdom and the United States. The European Commission takes part in the work of the OECD.

FOREWORD

Electricity fuels our increasingly industrialised and technology-driven economies and is a critical component of productive activity and daily convenience. Timely and sufficient investments in power generation are essential to ensure reliable supplies to customers. A diversified generation portfolio is required to be resilient to uncertainties. Investments must now also be cleaner, contributing to reducing the environmental impact of energy production. Policy makers are tackling the challenge of finding long-term solutions that efficiently balance the objectives of competitiveness, security of supply and environmental responsibility.

The current lack of long-term solutions creates great uncertainty about the regulatory and political framework, an outcome that is not conducive for investments. Uncertainty forces investors to be short sighted, creating a risk of under-investment. The risk is further compounded by growing electricity demand, ageing generation units and tighter environmental controls. It is now critical that existing resources are used as effectively as possible and that barriers to investment are reduced, first and foremost by reducing the considerable policy and regulatory uncertainty faced by investors today.

Against this background, governments of IEA member countries have raised concerns about the adequacy of investments in power generation, notably in the communiqué of the IEA Governing Board meeting at ministerial level in May 2005. This book responds to those concerns, which apply not only to electricity but to the entire energy sector. However, while many of the challenges in *e.g.* oil and gas relate to investment concerns in the relatively few resource rich countries, the adequacy of power generation is a matter close to home. The investment framework for electricity is determined by domestic decisions and policies. Governments can act to change the investment climate and action is urgently needed. This book outlines the areas where changes are most urgent, and recommends policy actions necessary to reduce investment barriers.

The strong inter-linkages with global energy markets pass uncertainties in the electricity sector on to other parts of the energy supply chain. Energy demand for electricity and heat generation constitute 40% of total OECD primary energy supply. Hence, the electricity sector is a large consumer of coal, natural gas, uranium, biomass and oil. Any investment uncertainty in power generation feeds through to uncertainty for those investing in the exploration, production and transport of coal, uranium and natural gas.

Four questions form the backbone of this book. *How large are the investment requirements in the short and medium term, and is there reason to doubt that the requirements will be met? What are the main decision parameters for investments in power generation? What are the critical elements of a framework that gives incentives for efficient investment responses? What are the main action points for governments and regulators in establishing a regulatory framework that facilitates proper investment?*

This book is the third in the IEA series on electricity market experience. It is closely related with a recent IEA publication on climate policy uncertainty and investment risk. It is published under my authority as Executive Director of the International Energy Agency.

Claude Mandil
Executive Director

ACKNOWLEDGEMENTS

The principal authors of this book are François Nguyen and Ulrik Stridbaek of the Energy Diversification Division, working under the direction of Ian Cronshaw, Head of Division, and Noé van Hulst, Director of the Office for Long-Term Co-operation and Policy Analysis.

This book benefited greatly from input, suggestions, comments and corrections from several contributors. Jolanka Fisher of the Country Studies Division and Fabien Roques of the Economic Analysis Division of the IEA, and William Blyth, Oxford Energy Associates, contributed with parts of the book and with comments. Professor David Newbery, University of Cambridge, gave important directions and comments. Considerable input was provided by Maria Argiri, Rebecca Gaghen, Brian Ricketts, Maria Sicilia-Salvadores, and Daniel Simmons of the IEA, Evelyne Bertel and Stan Gordelier, of the Nuclear Energy Agency, Doug Cooke, Australian Department of Industry, Tourism & Resources, Peter Fraser, Ontario Energy Board, Hanele Holtinen, Technical Research Centre of Finland, Malcolm Keay, Oxford Institute for Energy Studies, Karsten Neuhoff, University of Cambridge, John Paffenbarger, Constellation Energy, and Matti Supponen, European Commission. Members of the IEA Coal Industry Advisory Board offered their insight that led to review comments from Jorge Corrales, HC Energia, Mark Gray, AEP, Allan Jones, E.ON UK, Fernando Lasheras, Iberdrola, Kyohei Nakamura, J-Power, Shu Sakamoto, Tepco, Hans-Wilhelm Schiffer, RWE Power, and Juan Eduardo Vásquez, Endesa. Delegates of the IEA's Standing Group on Long-Term Co-operation gave valuable comments and guidance. Muriel Custodio managed the production of the book, Corinne Hayworth, designed the layout and the front cover, and Marilyn Smith edited the book.

TABLE OF CONTENTS

LIST OF TABLES

LIST OF FIGURES

LIST OF BOXES

EXECUTIVE SUMMARY AND POLICY RECOMMENDATIONS

In most IEA countries a new investment cycle in power generation is looming. A window of opportunity now exists to push for a cleaner and more efficient generation portfolio that will have significant impact on the energy sector and the environment for the next 40-50 years. However, the many uncertainties now inherent in the power sector create risks for investors, risks that may lead to under-investment – too little, too late, in the wrong location and with the wrong technology.

The recent liberalisation of markets delivers considerable benefits if implemented whole-heartedly and if backed by ongoing government commitment. In fact, competitive markets with cost-reflective prices are a strong, and most likely necessary, instrument to effectively balance energy systems in terms of economic efficiency, reliability and environmental responsibility. This ongoing process is, without question, one of the uncertainties for investors but its resulting risks can be greatly reduced when competitive and liquid markets are allowed to develop. The other most serious underlying uncertainties include CO_2 constraints, power plant licensing, acceptability of nuclear power, local opposition to new energy infrastructure, government support for specific generation technologies and government policies on energy efficiency.

Government action is urgently needed to significantly reduce this regulatory uncertainty. This would serve to establish effective competitive markets and provide firm policy directions in those areas in which markets fall short, such as taking account of environmental costs. Governments must also clarify and simplify power plant licensing procedures to accelerate the approval of new generation units.

A New Investment Cycle is Approaching ·····················

Margins of installed capacity over peak-load are often used as a measure of generation adequacy. These margins have recently decreased in several IEA countries, but are still comfortable in most regions. North America experienced strong capacity growth in 2000-04. Japan and Korea currently register high reserve margins, which continue to increase. Margins in Australia and New Zealand decreased, largely as a result of exploiting previous excess capacity. Europe has relatively constant margins, even though significant shares of new generation are from strong growth in wind power, which effectively decreases margins due to its low availability. Reserve margins are tighter in some countries and regions; a few are already facing a risk of capacity shortage within a few years.

Several factors will put more pressure on margins during the coming decade. Demand is increasing, particularly peak demand in several temperate countries. Existing power plants are ageing and environmental constraints are tightening. The IEA *World Energy Outlook 2006 (WEO 2006)* projects that installed capacity in the OECD will need to increase by 466 GW by 2015 – 20% of existing capacity. Most IEA countries have policy objectives to curb electricity demand through increased energy efficiency. This is often the most economical way to address climate change and energy security. By cutting demand it has the added benefit that fewer new generation plants are needed. If currently planned OECD policies on energy efficiency are implemented successfully, installed capacity is projected to increase by only 15% by 2015. Improved energy efficiency will ease the pressure on reserve margins. However, uncertainty about the effectiveness of energy efficiency measures is also a risk for today's investors.

It may be possible to delay demand increases. But the ageing of existing units, and eventual need for replacement, is inevitable. Most IEA countries experienced an investment boom in the 1970s, in response to the oil crisis. Many countries shifted generation portfolios away from oil, reinforcing the roles of coal and nuclear. Some of these units are now approaching the end of their lifetime. Of total installed coal, oil, gas and nuclear generation capacity in OECD countries, some 27% is now more than 30 years old. Investments in refurbishment and upgrades can extend the lifetime and capacity of some units, but this is a temporary measure.

Today, tightening environmental standards in IEA countries and the phase-out of nuclear power in some countries put even more pressure on the need for replacements. Most IEA countries have policies that set stricter environmental controls, particularly for coal-fired plants. For the oldest and smallest coal-fired units, it is more economical to decommission than to retrofit to meet new standards. In addition, some European countries have policies to phase out nuclear power. About 200 GW are projected in *WEO 2006* to need replacement in OECD by 2015. About one-third of installed capacity, equivalent to 872 GW, is projected to need replacement by 2030.

Making better use of existing power assets is one way to effectively delay the need for new generation capacity. Some IEA countries have liberalised their markets, comprehensively replacing regulated systems with frameworks in which competition creates incentives for efficient operation and investment.

Effective competition puts pressure on companies to use resources more efficiently and to practice just-in-time investment. This has allowed reserve margins to decrease without undermining quality in some pioneering markets including the United Kingdom, Australia, Texas and the Nordic countries.

Experience to date shows that, with the right incentives and with a stable investment climate, investors are responsive to the needs for new generation capacity. When signals are undistorted in effectively liberalised markets and companies have incentives to compete, investors respond to market signals and have so far added new capacity on time. New units – both large and small – are under construction and considerable capacity is planned across IEA countries. In addition, investors also seem to take the need for diversification into account when incentives are clear. With the current high natural gas prices, coal-fired generation is the cheapest option in many circumstances. A number of investors are building new coal-fired stations in several IEA countries and more are planned. Nuclear power is also often a competitive option when gas prices are high, and it is becoming even more attractive as CO_2 emission constraints tighten further. Nuclear power is again under serious consideration by investors in several IEA countries; the first decisions to make large capital investments have already been made. If this technology can break away from a history of delays and cost overruns, and overcome significant regulatory hurdles it is poised to halt and potentially even turn the current trend of decreasing shares of nuclear power in the OECD generation mix.

On a smaller capacity scale but with most noticeable growth, wind power capacity has become increasingly main stream in several countries, often backed by government subsidies such as feed-in tariffs. Combined-cycle gas turbines (CCGTs) are also being built in large volumes. The generation costs of CCGTs are sensitive to gas prices, but have several advantageous features: low investment costs, short construction time, modularity, relatively low CO_2 emissions and some operational flexibility. These features also make CCGTs, together with old coal fired units, ideal for mid-merit operation: they can operate fewer hours than base-load plants without undermining profitability. One important driver for large investments in CCGTs, particularly during the last decade, is that it has efficiently re-balanced generation portfolios that previously consisted primarily of traditional base-load plants.

All in all, generation units that are now planned or currently under construction will increase capacity to approximately meet that portion of the projected gap that corresponds with increasing demand by 2015. More new capacity will be needed to replace decommissioned units, even if improved energy efficiency helps to counterbalance overall supply and demand.

There is no reason to doubt that investors understand the need for large investments; indeed, significant additional capacity is already planned. However, a large portion of the needed generation projects is still awaiting final investment decisions that must be made in the coming few years. Such large decisions will be taken on time only if investors can clearly see that expected returns will outweigh expected costs – even when accounting for uncertainty and risk.

One of the most difficult decisions for investors is the choice of technology, which ultimately determines implications for the environment and for security of supply. There are no clearly superior technologies among the group of more mature generation options. Moreover, a well-diversified generation portfolio, designed to deliver supply efficiently both now and in the future, will have to include several technologies. Thus, the choice for investors depends on many factors and is always made with an eye on potential for profit. Small changes in the key cost factors (*e.g.* investment costs, fuel costs, CO_2 emission costs and utilisation rates) can completely change the relative ranking of technologies in terms of total generation costs levelised over the lifetime of the plant. Well-functioning markets for electricity, fuel and CO_2 emissions provide strong incentives for investors to diversify and to opt for cleaner technologies although diversification is, obviously, limited to the technology options actually available. Governments play a critical role in keeping as many options open as possible by supporting R&D of new technologies and through effective regulation and policies, particularly in the case of nuclear. In fact, nuclear power will only become more important if governments in countries where nuclear power is accepted play a stronger role in facilitating private investment, especially in liberalised markets.

Timing of investment is another key challenge for investors. Lack of strong and clear incentives, especially when coupled with high regulatory uncertainty, could result in under-investment. Projects with higher risks are more costly: thus, investors must be relatively certain of realising a higher rate of return on capital. In an environment of significant uncertainty, there is a value in delaying costly investments until new information or new technologies reduce the degree of uncertainty and the associated risk. Given the long lead times and often significant capital requirement in building new generation facilities, government action to reduce regulatory uncertainty – and thereby to help lower and manage investment risks – is urgently needed in the very near future. Without such action, the perceived risk level of the current investment environment may undermine the incentives for making investments in projects with large up-front investment costs, such as nuclear. Investors will place too much emphasis on short-term factors. The financial and operational flexibility of CCGTs could effectively turn this technology into the default option for investors. Even worse, lack of government action may delay investment altogether, putting efficiency and reliability seriously at stake, and incurring unacceptable environmental consequences.

Key Message

Governments must ensure a stable and competitive investment framework that sufficiently rewards adequate investments in a timely manner.

Considerable investment in new power generation will be required over the next decade to meet increasing demand and replace ageing generation units. Current trends suggest a significant risk of under-investment. Long project lead times and high investment costs, particularly for large base-load units, create a need for government action to reduce uncertainty in the very near term. Efficient use of existing resources is particularly important at this stage, as it allows for lower margins and buys time to meet investment requirements.

Pricing CO_2 Emissions can Drive Clean Investments ....

It is increasingly clear that action must be taken to reduce greenhouse gas emissions and lessen the impact of climate change. The question of *what* actions are needed is far from resolved. Uncertainty about the future direction of policies under development in these areas represents a significant disincentive for current investment in power generation. Continued lack of clarity may lead to intolerable outcomes for electricity systems. Credible, long-term policy commitments can critically reduce risks of making – and indeed encourage movement towards – investments in a cleaner and more efficient generation mix that will greatly influence the environmental footprint for the next 40-50 years.

Governments are best positioned to assess, on a broad scale, the environmental risks and costs associated with power generation, and possible macro-economic implications resulting from too high dependence on, for example, natural gas imports. That said, governments are not necessarily best equipped to actually manage risks by picking preferred technologies and generation portfolios. Many clean and non-import dependent technologies, such as some renewable technologies, and nuclear power, need government backing that reflects the added benefits for the environment and from reduced import dependence. Commercial investors have a long history of managing risk in the marketplace and are best placed to assess the optimal choice and combination of technologies, taking into account technology maturity and efficiency concerns. Governments and commercial investors are complementary. The principal role of government is, through market-based instruments, to create incentives for investment decisions that support policy objectives on environment and security. In contrast, a command-and-control approach to policy making, that seeks to dictate investment, can lose sight of the efficiency dimension and render competition irrelevant as an instrument to signal investment needs. Too much intervention from the policy side can effectively re-introduce a fully regulated system.

Market-based instruments are already available for several environmental policy objectives; they have shown the potential to improve cost effectiveness and are compatible with liberalised electricity markets. Putting a price on CO_2 through taxes and CO_2 emission trading schemes introduces strong incentives to which investors are already reacting. Support of clean technologies through quota-based trading systems (such as renewable portfolio standards and tradeable green certificates) shows potential for broader application, particularly in that they also ensure transparency, efficiency and market compatibility. However, they are still relatively new. Important lessons, based on practical experience, still need to be learnt about such instruments. Tax credits and other similar financial premiums, which leave many incentives for efficiency intact, may also ensure transparency and market compatibility, even though they may remove some of the competitive pressure. Direct financial support systems that protect investors from most risks (*e.g.* feed-in-tariff systems) may be useful for development of new technologies. However, they create relatively weak incentives for investors to consider both the costs of integrating new technologies into the larger electricity system and the fundamental resource conditions in the market. It is also more difficult to ensure necessary cost transparency, efficiency and market compatibility with such direct support systems.

The European Union emission trading scheme (EU ETS) is a promising instrument to reduce uncertainty and achieve cost-effective emission reductions. In putting a price on CO_2 emissions, it transforms a policy goal (emissions reduction) into a quantifiable cost factor that investors can take into account when making decisions. But the EU ETS only runs until 2012, in line with the Kyoto Protocol. Such a short timeframe actually increases investment risks and costs, and limits options for investors. Investment decisions taken today are already geared toward units that will be commissioned close to the initial 2012 limit for the EU ETS. Thus, investors do not know whether the policy will remain in effect and cannot accurately account for its consequences. In addition, several EU member states have shown a lack of commitment to use the EU ETS for actual emission reductions at least as currently structured. For example, some governments have issued free allowances to existing and new generators in ways that effectively turns the EU ETS more into a tool to encourage portfolio diversification. This undermines the functioning of electricity and emissions markets, and also puts efficiency and environmental objectives at stake. Adjustments of the design of the EU ETS, including allocation and coverage, will be necessary to increase the effectiveness of the EU ETS to reduce CO_2 emissions in the future.

IEA countries generally give subsidies to specific renewable technologies; wind power is the most prominent recipient of such support. Some countries already have considerable shares of wind power, and total installed capacity is increasing rapidly. On-shore wind power in favourable locations is moving into the ranks of more conventional technologies, and putting a price on CO_2 may soon make specific additional subsidies unnecessary. The strong development of wind power capacity now raises important concerns regarding its integration into electricity systems. Wind power can only be generated when there is sufficient wind. Thus, back-up resources must be available when the wind stops blowing. This has implications for the real-time operation and balancing of the electricity system, as well as for the total costs and the long-term development of the generation portfolio and the transmission system.

All generation technologies incur certain integration costs, which depend greatly on the way trade is organised. Poorly designed electricity markets that mainly facilitate trade for large generation units unnecessarily increase integration costs, particularly for wind power. So far, it has been possible to integrate relatively large shares of wind power – at acceptable costs – in systems that have strong interconnections with a larger wind-poor system. However, integration costs will rise if the same or even higher shares are integrated uniformly across entire systems. Integration costs of high uniform shares of wind power are understood in theory, but are untested in practice. It remains unknown what total shares of wind power an interconnected electricity system can support without incurring unacceptable integration costs and without jeopardising system security. It is important that wind power is not chosen by governments as a winning technology under all circumstances, but rather that investors be given incentives to carefully consider integration costs. This will leave investors with incentives to compose a well-diversified generation portfolio.

Key Message

Governments urgently need to reduce investment risks by giving firmer and more long-term direction on climate change abatement policies.

Putting a price on greenhouse gas emissions is an effective way to internalise the costs of climate change. Direct financial support for specific technologies, such as renewables and nuclear, should be done at the lowest cost and with market-compatible instruments. Market-based instruments, such as tradeable obligations systems, have many advantages; direct subsidies, such as tax credits, can also be implemented in ways that are compatible with competitive markets. Nuclear power will only play a more important role in climate change abatement if governments in countries where nuclear power is accepted play a stronger role in facilitating private investment.

Competition is an Effective and Necessary Tool ....

Liberalisation of electricity markets was introduced in several pioneering markets with considerable success to facilitate more efficient use of resources. In previously regulated markets, there was a tendency to over-build and to shift all risks and costs directly to consumers. Liberalised markets force competing investors to properly account for all risks and have removed incentives to over-build. This correction is one of the most important positive effects arising from the introduction of competition. Most IEA countries are now implementing regulatory reforms that have introduced competition at varying speeds. But liberalisation is not implemented – or embraced – in a single swift operation. It is a process that requires whole-hearted and committed implementation, and ongoing government backing.

In an initial phase, competitive markets delivered short-term operational efficiency improvements from optimised dispatch across large areas. With increased use of existing resources as an alternative to new generation, reserve margins have now decreased to levels at which new investment is needed in several of the pioneering markets. In this second phase, liberalised markets need to demonstrate that they can also provide efficient incentives for investments and deliver benefits in the long term.

The performance of liberalised markets as a tool for investment in power generation is a particularly contested area of electricity sector reform. Early experiences indicate that competing investors are responding to market signals. Tighter supply, increasing natural gas prices and environmental constraints are directly reflected in increasing wholesale electricity prices, particularly in those markets in which electricity prices are competitive and cost reflective. Investors are responding by investing in new generating capacity. So far, investments were concentrated on gas-fired generation. A new nuclear power reactor is being built, benefitting from unique contracting arrangements between utilities and large consumers. Coal-fired plants are proposed in several countries.

Incentives from competition improve the use of capital, mainly through more effective risk management, which ultimately reduces costs. Risks that were passed on directly to consumers in regulated systems have become transparent in competitive markets, and are directly reflected in wholesale electricity prices. Moving from a regulated system where consumers are forced to pay the bill does, however, not imply that consumers should not commit to covering the costs of generators. The costs of managing investment risks can be reduced considerably if producers and consumers co-operate. Generators and consumers represent a wide range of interests with different risk profiles. Both share, at least to a certain extent, an interest in stable prices to manage risks. Moreover, if longer term commitments can reduce costs for investors, consumers will benefit by entering into longer term contracts. Thus, trading and contracting arrangements become critical for investment costs, while also offering consumers effective protection from price volatility. In several markets, liquidity in traded contract markets has increased considerably. This allows for dynamic, efficient and low-cost risk management and offers an opportunity for independent generators and retailers to operate in the market, which also improves competition.

Some governments offer regulated prices to certain groups, as an alternative to competing offers based on wholesale prices. While regulated prices are offered to protect consumers from risks, the practice has been shown to severely distort competition. Measures to regulate retail prices may be necessary in a transitional phase until effective competition has developed, but it is doubtful that such measures serves consumers interests best in the long term; the more effective alternative is to protect consumers by ensuring strong competition.

Key Message

Governments should pursue the benefits of competitive markets to allow for more efficient and more transparent management of investment risks.

Competition in well-designed and effectively liberalised markets creates incentives for efficient use of resources and investments in power generation. However, in order to deliver its anticipated benefits, liberalisation requires whole-hearted implementation and long-term commitment by governments. Competition cannot always stand alone. When necessary, governments should pursue intervention in ways that complement the market and facilitate its functioning.

Independent Regulators and System Operators Make Competition Work ....

In order for liberalised markets to deliver the intended benefits, competition must flourish across all levels. The danger of a concentrated market is that firms with market power may not have sufficient incentives to invest. Indeed, withholding new investment could be a means for dominating firms to push up prices and increase profits – outcomes that are ultimately detrimental to public welfare. Such a strategy can succeed for extended periods only if dominating firms can, at the same time, block or obstruct investments by competing firms. Thus, it is important to create the right conditions to encourage competing firms to enter markets, including rules and market design that are clear, efficient, and ensure equal treatment for all players. To this end, independent regulators and independent transmission system operators play critical roles in establishing trading rules and ensuring fair access to networks. These roles must be effectively separated from generation and retail supply.

Trade and co-operation across jurisdictional borders also creates important benefits of liberalisation. Resources are used more efficiently, which allows co-operating systems to operate reliably with lower reserve margins. Cross-border trade is constrained by available transmission capacity, but with an appropriate market design the benefits also create incentives for investment in new transmission interconnections. The benefits are even more significant for smaller systems; indeed for smaller markets, cross-border trade may be the only way to improve competition amongst local generators. Once initiated, competition must be allowed to drive the organisation of the sector to deliver the full efficiency potential – even if this requires larger, consolidated firms. If it is not possible or desirable to break up dominating firms, the only other option may be to enlarge markets through integration.

Effective cross-border trade requires extended regulatory harmonisation across interconnected markets. Several countries, such as the United States and Australia, have successfully opted for models in which a single regulator covers all integrated markets – at least in matters directly related to cross-border trade. These are interesting examples for the EU, which still lacks a single regulator for cross-border trade.

Shrinking reserve margins and increasing cross-border trade pose new challenges for system operation. A comprehensive legal framework must effectively allocate responsibilities and requirements for secure system operation. This must be backed with effective regulation and seamless co-operation amongst interconnected system operators. With the right framework, cross-border trade provides opportunities for more efficient and reliable electricity systems. Without such a framework, the potential benefits can quickly be transformed into a threat to system security.

Key Message

Governments need to ensure that independent regulators and system operators establish transparent market rules that are clear, coherent and fair.

Transmission system operators hold the key to competitive electricity markets and must be effectively separated from generation and retail supply. Unified regulation and unified system operation should be pursued as tools to facilitate dynamic trade across borders and efficient sharing of reserves.

Cost-reflective Prices Vital to Adequate Investments........

Amongst the market designs that have effectively created market signals for investors, two areas stand out as particularly important and challenging: locational pricing and pricing of scarcity. Price clearing mechanisms that give sufficiently strong locational signals can help to ensure that investment in new generation is strategically placed - *i.e.* in areas where it is most needed or where energy sources are abundant. Locational price signals are also crucial for investment in new transmission capacity. In markets with weak locational signals, poor signals to investors often trigger a need for interventions. If the coming investment cycle is allowed to unfold without appropriate locational signals to investors, such distortions are likely to increase.

The other critically important feature is the need to allow prices to increase to levels that reflect full costs, particularly during times of scarcity. Price caps on wholesale prices or regulated retail prices can critically distort market outcomes. When incentives for investment in peak-load resources are capped, reliability is jeopardised. Price caps and other price distortions also undermine the long-term confidence in the market, effectively distorting all investments.

In the initial phases of market reform, caps on wholesale and retail prices were deemed necessary in some cases to curb abuse of market power. However, other markets have shown that it is possible to curb systemic abuse of market power without the use of low price caps. Powerful competition legislation, strong competition regulators and effective co-operation between electricity and competition regulators are prerequisites for effective competition regulation. Market screening methods and regulatory measures have both been applied in efforts to mitigate market abuse and experience with them is accumulating. Regular and comprehensive market screening studies are conducted in several US markets. Several competition regulators, including the German regulator, have recently conducted in-depth competition inquiries.

The lack of participation from the demand side is an important justification for the need to reduce market power during scarcity. Consumer response to high prices – typically by shifting demand to other products or to other time segments – is an integral element in markets for most other goods and services. This feature is still largely lacking in electricity markets, which puts the supply side in a strong position. There is no reason to doubt that electricity demand is price responsive in principle. Different uses of electricity obviously have different values to different consumers. Transaction costs are the barrier in the electricity sector: they are currently too high for most consumers to easily express their preferences. But in reality, even a very small degree of demand response can play a critically important role for system balancing during periods of scarcity. Certain limited volumes of demand have responded to price spikes in markets that allow prices to reflect scarcity. The potential for demand to actively participate in electricity markets is significantly larger than has been realised to date. Better metering of more consumers is an important step for accessing this untapped potential. Several IEA countries are pursuing projects to provide smart meters to all consumers. Italy is now in the last phase of a full roll-out of 30 million meters to Italian households.

When low price caps are used, incentives for investment in power generation are capped with it. In the United States price caps have necessitated additional incentives by imposing capacity obligations and capacity markets. This approach has been fraught with problems that regular market design modifications are continuing to try to improve. Ineffective design of capacity markets creates a tendency for markets to revert back to the over-building that was characteristic of previously regulated regimes. Price caps and specific design features in capacity markets also tend to distort incentives for demand response, which is a critical shortcoming in the first place. It has also proven difficult to properly account for the potential of cross-border trade, thereby missing some of the key benefits of competition. So far, capacity markets have been a poorly prescribed medicine with serious negative side effects. With low price caps and muted price signals, capacity markets may be necessary. However, it would be more effective to remove price caps than to try to administratively repair their consequences.

Intervention is deemed necessary to secure reliability during the transitional phase in some markets, while competition and demand response are developing. In fact, some type of soft intervention is evident in most markets. Strategies to focus interventions on facilitating demand response without muting price signals have proven useful in the transition towards robust markets. Such transitional capacity measures must be based on reliability criteria that are adapted to the new competitive framework – not on criteria associated with the old, and often over-built, regulated systems.

Key Message

Governments must refrain from price caps and other distorting market interventions.

Wholesale electricity prices are inherently volatile and price spikes are an integral part of a competitive market. Price caps, regulated tariffs that undercut market prices, and direct market intervention seriously undermine market confidence, jeopardising efficiency and reliability. Governments can best address systemic market power abuse by: improving market design; strengthening competition law and competition regulators; and diluting the dominance of large players. Demand response constitutes an essential but still poorly exploited resource, and must receive specific attention in the development of market design and regulation. Increased installation of better metering and control equipment could considerably strengthen potential demand response resources. Capacity measures may be necessary if price caps are imposed. However, they are not a preferred solution to address market power and can easily become a barrier for the development of a robust market.

Delays Frustrate Markets

All of the preceding factors – good market design, effective regulation, competition and clear, long-term environmental policy – matter little if investors cannot obtain permission to build power plants and transmission lines. Governments have a responsibility to balance protection of private property rights, public welfare, and the local and global environment. A very worrying tendency amongst the general public is to react to new investment with a "not in my back yard" (NIMBY) attitude. These responses reflect increased environmental concern, both on a local and global level, and increased value of property.

During the past phase where new investments in generation and transmission were less important, these trends were manageable. Now that the need for considerable new investment is evident, a re-balancing of interests is essential. Governments urgently need to establish clearer, shorter, more integrated and more comprehensive application procedures for new plants. The responsibility to balance interests should remain with governments, rather than investors. One way to improve licensing and approval procedures is to reduce the number of approval bodies and phases. Ideally, investors should have access to "one-stop-shop" licensing, in which one official body holds as many of the approval responsibilities as possible or at least is given the duty to co-ordinate. Italy's efforts to streamline and shorten licensing, and to concentrate authority to one body, have proven effective in accelerating investments in power generation. Locational pricing can also provide some incentives to local communities to accept new infrastructure.

Higher natural gas prices and increasing CO_2 emission constraints have improved the competitiveness of nuclear power, in part by emphasising this technology's benefits in terms of security of supply and climate change abatement. Viability of nuclear power is highly dependent on the regulatory framework. Thus, for countries that accept nuclear power and want to keep that option open, effective plant licensing is vitally important. Uncertainty about design and plant licensing, as well as the costs of waste management and decommissioning, are all sufficient to undermine the technology's overall cost competitiveness. Every year of delay in finalising an ongoing nuclear project adds 3% to the total costs of generating electricity. There is a significant scope for international co-operation to streamline plant licensing procedures.

Government intervention to facilitate further development of nuclear technology is not enough. It must be accompanied by broad public acceptance and governments have an important role to nurture the necessary public debate. To that end, firm decisions about radioactive waste management and decommissioning have proven to be critical.

Key Message

Governments must implement clearer and more efficient procedures for approval of new electricity infrastructure.

Delays caused by slow licensing and inefficient approval procedures frustrate markets, and are serious barriers to timely investment. Governments must re-balance competing interests in favour of new electricity system infrastructure and offer clearer and more efficient approval procedures, preferably centred on one approval body. Timelines for approval processes must be clear and established in advance. Fast and efficient licensing is particularly important for new nuclear power plants which face very high risks as well as long planning and approval process. Early public debate is essential for the acceptance of necessary new infrastructure.

INTRODUCTION

Across IEA countries, ageing power plants and growing energy demand are the primary drivers behind an increasingly urgent need for considerable investment in power generation in the coming decade. There is no reason to doubt that the necessary investment resources are available in these developed countries. The greater challenge is to attract the resources available to the energy sector.

At present, various aspects of the sector itself present serious barriers to the development of adequate volumes of new capacity: complicated and time-consuming processes for licensing and approval, uncertainty created by the recent and ongoing transition to liberalised markets, and the development of new technologies that change the overall energy mix. At the same time, there is a notable lack of clarity in the policy realm, particularly in relation to efforts to mitigate the environmental impact of energy production and consumption. Together these factors may make energy less attractive to investors than other market sectors.

Balancing efficiency, reliability and environmental responsibility requires not only quantity but also quality of investment. This book explores each of the above areas in relation to the overall objective of ensuring that investments will be made in the right locations, at the right time, in the right quantities and with the right technologies. Governments can play a critical role by providing strong incentives for investors, essentially creating the right signals to avoid under-investment, but without triggering a response that leads to expensive and unnecessary over-investment.

New generation capacity is an important aspect of meeting increasing energy demand, and is the main focus of this book. However, new capacity is not the only option. Increasing demand can most effectively be met through a mix of sources: new generation units, better use of existing units, imports using transmission infrastructure, and even reductions of other types of demand. Investment in transmission systems and better integration of consumer preferences are thus important alternatives to new generation resources and investment in these alternative resources also creates considerable challenges for policy. Under current conditions, obtaining permission to build new transmission is at least as difficult as getting approval for new generation. This makes it even more important to ensure that existing transmission lines are used well. On the demand side, many current problems related to efficiency and reliability originate from the existing disconnect between supply and demand. The electricity sector was traditionally overly focused on the supply side. Competition and communication technologies now make it possible better to account for consumer preferences. For example, when consumers respond to price by shifting demand to other time periods, resources are released to be used for other purposes.

The topics of transmission systems and demand participation cover a wide range of opportunities and challenges that merit separate detailed analysis in future IEA work. Since they cannot be separated completely from the generation side, they are given some emphasis throughout this book. A recent IEA publication, *Lessons from Liberalised Electricity Markets,* provides an overview of generation, transmission and the role of electricity consumers in the context of examining best practices in pioneering liberalised markets (IEA, 2005a).

Liberalisation of electricity markets is important to this current discussion in that it fundamentally changes the incentives for investment in new generation capacity and for utilisation of existing capacity. Competition is being introduced in all IEA member countries, at varying speeds. Some pioneering markets began comprehensive reforms more than a decade ago and have now achieved relatively advanced stages with considerable success. These markets include the United Kingdom, the Nordic countries, Australia, New Zealand, the state of Texas (US) and several north-eastern US states, and Alberta (Canada). Liberalisation has also been implemented with varying degrees of success in several non-IEA member countries, starting with Chile in the 1980s.

This book focuses on adequacy of generation capacity to deliver electricity to consumers – in the quantity and quality desired, at the least possible cost. Generation adequacy is, however, only one element in a supply chain that starts with fuel input and ends with reliable electricity supply. Quality can vary slightly, but outside a relatively narrow quality range the delivered electricity cannot be used without damaging appliances and machines. For the consumer, it matters little where the chain breaks down in the case of interrupted supply. But to properly analyse electricity system performance and regulation, each link in the supply chain must be examined separately.

Analysis shows that in liberalised markets – *i.e.* where previously vertically integrated utilities have been unbundled into generation, transmission, distribution and retail supply entities – each link of the supply chain must be regulated separately, and with tailored, targeted objectives. In an unbundled sector, four main links support reliable electricity supply: upstream energy (fuel) security, adequacy of generation assets, adequacy of networks, and system security (secure real-time system operation).

This book also discusses the regulatory framework required to provide competing market players with incentives to invest in generation capacity. Adequacy of generation is, however, also inextricably linked to other elements of the supply chain. In fact, it is often difficult to make clear-cut distinctions between them.

The interface between generation adequacy on the one hand and transmission adequacy and system security on the other is an interface between a strictly regulated world and a competitive market. Transmission system adequacy was expected to develop from incentives in competitive markets. In reality, it has proven difficult to establish a sufficiently firm and clear competitive market framework to deliver in this area. To date, the role of transmission systems remains deeply integrated in reliability criteria; investment in competitive transmission – i.e. merchant lines – is still a rare exception rather than the norm. Transmission systems still rely heavily on regulatory intervention, and are often substitutes for generation. The case is the same for distribution systems. In fact, failure of distribution systems is, by far, the most common cause of supply interruptions. System security also requires direct regulatory intervention. Independent system operators must have clear responsibility to keep the lights on in real-time system operation, and they must be effectively bound by clear reliability criteria.

At the other end of the supply chain, security of fuel supply is an issue receiving great political attention, potentially with great impact on investment in generation and the choice of generation technologies by competing investors. Commercial market players have incentives to diversify as part of their risk-hedging strategy. Variations in fuel prices affect plant profitability directly so expected fuel prices and probabilities of interruptions in fuel supply are important parameters in investment decisions. The profitability of a diversified generation portfolio is more resilient to fuel price hikes or interruptions. The challenge is that energy security has impacts on two different levels. Commercial players have a narrow focus on potential profits and losses at individual plants. In contrast, governments need to consider the aggregate effect of generation being overly dependent on monopolised fuel import markets, which can have an impact on national economies. At present, the difference between these micro- and macro-economic effects is not well documented or well understood. Thus, it is not obvious how governments and regulators can intervene to adjust incentives to encourage diversity in a way that also unambiguously improves outcomes in terms of a better diversified generation mix at a cost that can be justified – even when the real costs of risk are taken into account.

One of the key aims of competition in liberalised markets is to create incentives for adequate investment in power generation. Lessons from liberalised electricity markets (IEA, 2005a) highlight that incentives created through competition rely on effective market design and that it is critically important to implement liberalisation comprehensively without undue government intervention. Under the previously regulated sector, incentives to invest in new power generation reflected the quality of the regulatory regime. Regulatory approaches varied greatly across IEA countries and only a few countries, such as the United States, had formal independent regulatory bodies. A common feature was that regulators or other authorities represented the consumer when determining investment requirements and approving electricity tariffs. At the same time, investment adequacy was assessed based on a set of reliability criteria – including acceptable margins of installed generation capacity over peak-load. Minimum acceptable reserve margins and other reliability criteria also varied from system to system, depending on factors such as the shares of hydro power and the demand profile. Acceptable reserve margins were often in the 20-30% range.

In a regulated system, investment risks are effectively passed directly to the consumer. This can work well if regulators and authorities with regulatory functions truly represent the preferences of consumers and effectively put pressure on utilities to increase efficiency. In reality, regulatory regimes tended to lead to over-building and it was difficult to maintain sufficient information to ensure efficiency. Cramton and Stoft (2006) calculate that reserve margins at the traditional regulated levels in the United States correspond to reliability values that far exceed the preferences indicated in consumer studies. In other words, the high reserve margins were effectively forcing consumers to buy something that they did not want. When the rate of demand growth slowed, the negative effects of over-building were aggravated further. With the emergence of technologies that support advanced communication and trade, liberalisation was perceived as an effective way to improve efficiency in the electricity sector – as it had been in several other traditionally public sectors.

With effective competition in liberalised electricity markets, it is no longer guaranteed that investors can pass on all risks to consumers. Power generators have to compete with other generators when establishing contracts with retail suppliers and consumers. It may be possible to negotiate contracts that cover many of the risks, but poor investment decisions are likely to result in a loss for the investor. In liberalised markets, power generators have incentives to put existing resources to the best use and to make just-in-time investments. A natural consequence is that reserve margins are lower. Falling

reserve margins in liberalised markets are, thus, to be expected and a clear sign of successful liberalisation. Increased trade and co-operation also allow reserve margins to decrease. If this is managed well by regulators and system operators, the benefit is unambiguous – the same level of quality is delivered with fewer resources. Falling reserve margins in liberalised markets can also be attributed to the fact that generation adequacy is better aligned with the level of reliability that consumers actually prefer and are willing to pay for.

Regulators and system operators play important roles in terms of establishing checks and balances to ensure that reliability does not fall below acceptable levels. However, analysis of adequacy must be adapted to the new competitive framework. This adaptation remains incomplete in many jurisdictions; generation adequacy is still assessed according to similar criteria used in the previously regulated regime – despite the fact that competition has fundamentally changed the framework (Joskow, 2006). A recent study in the United Kingdom found that a 5-10% reserve margin is adequate to maintain reliability within the UK system, but also points out that reserve margins provide a very crude measure that will change with the generation mix (OXERA, 2005). In Australia's National Electricity Market, reserve requirements of available capacity to meet reliability criteria were less than 5% above peak-load in 2004.

More effective use of resources is a key driver behind liberalisation – and an important means of reducing the need for expensive new investment. The critical issue in liberalised markets is whether incentives are in place to ignite new investments when they are actually needed – which they will be eventually.

Thus, a key question addressed in this book is how to create the right environment for future investment in power generation. This encompasses two interrelated areas: providing incentives for new investment and creating frameworks that support effective risk management. In liberalised markets, incentives are linked to the price of electricity and risks are managed through contracting, effectively lowering investment costs.

With price and contracting as its main themes, this book draws considerably on previous IEA studies. It is the third book in a series on energy market experiences. The first, *Lessons from Liberalised Electricity Markets* (IEA, 2005a) outlines the main issues that have arisen with power market reform. The second, *Learning from the Blackouts* (IEA, 2005b), focuses on electricity system security in competitive markets. This series was preceded with a series on electricity market reform, which included *Power Generation Investment in Electricity Markets* (IEA, 2003) and *Security of Supply in Electricity Markets*

(IEA, 2003), both of which examined the principles of investment adequacy in competitive markets, thus constituting key reference points for analysis in this present book.

Historical developments and trends in terms of installed capacity, generation and demand in IEA countries are discussed in Chapter 1. After outlining current physical challenges, Chapter 2 moves on to study key decision parameters for investors, including costs and uncertainties, incentives to diversify and particular factors related to integrating the rapidly expanding wind power sector. Chapter 3 focuses on the actual decision-making process in a competitive market framework. In addition to exploring incentives for adequate investment, it discusses the role of cost-reflective prices and the role of additional regulated incentives such as capacity markets. Chapter 4 assesses some of the key roles and responsibilities for governments and regulators in establishing effective markets, including facilitating the approval and construction of new generation plants and addressing the challenges associated with nuclear projects.

STATUS AND TRENDS IN POWER GENERATION

The power sector in the OECD experienced profound changes in the last decade. Although these changes differ in magnitude among countries, some common trends and issues are evident. Overall, the OECD moved towards consolidation and market liberalisation, resulting in more competitive electricity markets. Institutional and regulatory frameworks were adapted to reflect evolving market rules and structural changes. The search for ways to reduce harmful emissions imposed new constraints and created new opportunities for generators and investors. Some jurisdictions experienced mixed results and concerns about market power are still prevalent.

Not surprisingly, this period of evolution created a sense of uncertainty – and indeed vulnerability – across the power sector. Market uncertainty was amplified by the surge in fossil-fuel prices and the increased volatility of both fuel and electricity prices. Although not due to any lack of generation capacity, major blackouts in North America (2003) and Europe (2003 and 2006) revealed the vulnerability of interconnected power systems (see Box 1.1).

During this time of transition, such uncertainties affected investment in power generation. Rapid change made it difficult for investors to determine when to invest, where to locate new generation facilities and even what technologies to invest in. Ten years on, it is increasingly evident that energy demand will continue to rise over the same time period in which a large number of existing plants will reach the end of their useful lives. Major decisions about power generation must be taken in the very near future. Moreover, the decisions must be taken within a context in which policy makers and the general public demonstrate increasing concern about the environment.

> ### Box 1.1 . Generation adequacy and blackouts
>
> *Large-scale blackouts of electricity systems hit North America, Italy and Scandinavia in 2003*. Another serious disturbance blacked out a large portion of the European electricity system in 2006±.*
>
> *These events had somewhat different causes. However, subsequent analysis reveals a number of parallels and key lessons to be learnt for policy makers and regulators. One commonality is that none of the blackouts occurred during times when load was at peak levels. In fact,*

the Italian and cross-European incidents happened during the evening and night on a weekend. These blackouts were not caused by a lack of generation capacity to meet peak-load.

The most critical common lesson from all the incidents is that principles for reliable system operation must change and adapt to the new reality of liberalised markets, where cross-border trade plays a much larger role. The United States responded by taking numerous measures to reinforce the regulatory framework within the Energy Policy Act 2005. Today, reliability criteria are legally enforced. In Europe, the European Regulators Group for Electricity and Gas (ERGEG) is now proposing similar measures.

* The events in 2003 have been thoroughly researched by involved regulators, system operators and reliability organisations. Key findings and policy recommendations are explored in Learning from the Blackouts: Transmission System Security in Competitive Markets (IEA, 2005b).

± Preliminary findings and conclusions for the European events in 2006 have been published by the Union for the Co-ordination of Transmission of Electricity (UCTE, 2006), and the European Regulators Group for Electricity and Gas (ERGEG, 2006).

Against this background, investments in power generation[1] have continued to expand in most OECD countries over the last 15 years. IEA statistics show that for the OECD as a whole, total installed generating capacity increased from 1 715 GW in 1990 to 2 400 GW in 2004; a total increase of 685 GW, or approximately 49 GW annually. This represents an average rate of growth of 2.4%.

Over the same period, total OECD electricity consumption increased at an average annual growth rate of 2.3%. Thus, at the aggregate level, one might observe that capacity kept pace with demand. However, since electricity markets are regional in nature, it is more appropriate to assess generation adequacy at the regional level. A closer look at national electricity statistics reveals that the investment picture varies considerably between countries.

In the last decade, most countries concentrated their power generation investments in gas-fired power plants, especially combined-cycle gas turbines (CCGT), (Figure 1.1). For the 1990-2004 period, gas-fired capacity in the OECD countries quintupled from 76 GW in 1990 to 380 GW in 2004, increasing at an average rate of 12% per year. However, in recent years, growth has slowed significantly for several reasons, including higher and more volatile gas prices and surplus gas-fired capacity in some regions. A shift towards more coal-fired generation is also in the horison. At the same time, concerns

---

1. In the context of this report, investments in power generation are measured by increases in installed capacity.

about rising CO_2 emissions have accelerated investments in renewables and revived interest in nuclear (in countries in which nuclear is acceptable). From 1990 to 2004, nuclear capacity in the OECD increased at an average rate of 1.2% per year (up-grades of existing units played an important role) while hydro capacity increased by 1% and coal-fired capacity by 0.8%.

Figure 1.1

Net changes in OECD generation capacity
(units installed, under construction and planned, 1980-2015)
show shift from gas towards coal

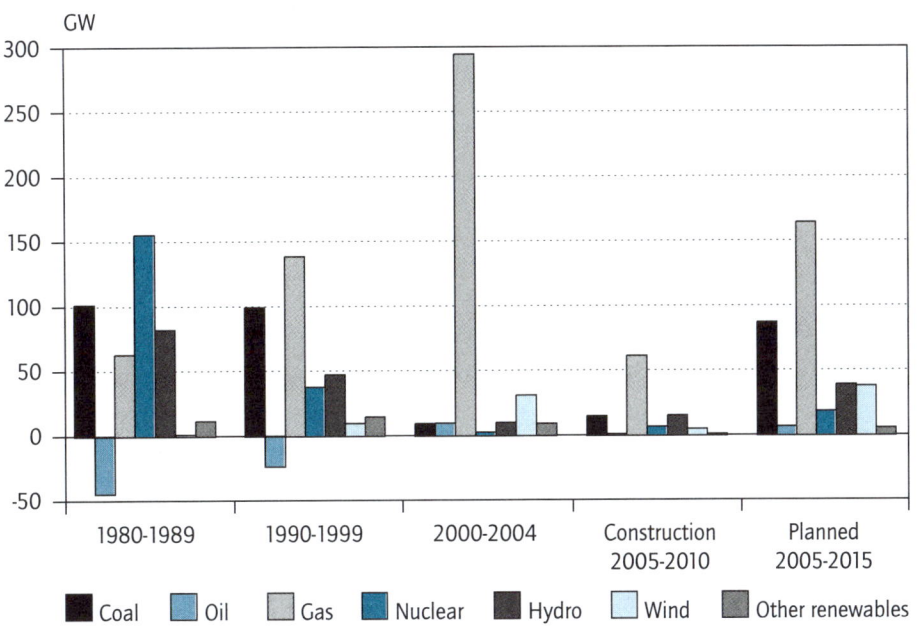

Sources: IEA statistics; Platts 2005.

Figure 1.2 illustrates the fuel sources used to meet increasing demand in the OECD over the past five years. Gas is fueling a large share of the increasing demand but in close interaction with other sources but the new gas-fired generation capacity is not used at full capacity. The utilisation of particularly coal-fired capacity has increased, meeting large shares of increased demand with the same level of installed capacity.

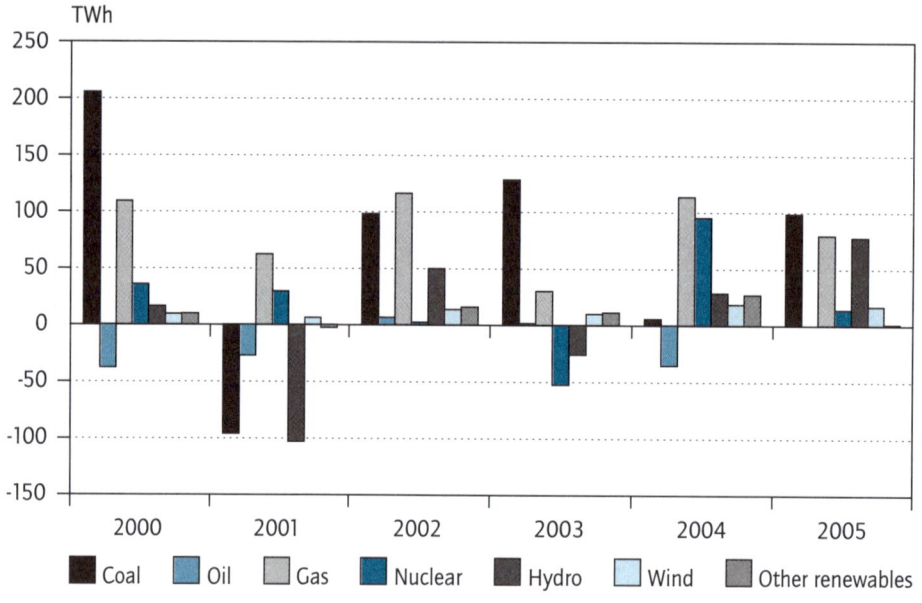

Figure 1.2

Year-on-year change in generated electricity shows gas-fired generation as main contributor, in dynamic interaction with coal-fired generation, in OECD, 2000-05

Source: IEA Statistics.

In many countries, the push for renewables resulted in considerable capacity additions in wind power. Other sources of renewable energy such as biomass, small hydro and solar have also experienced varying degrees of growth. Wind power capacity in the OECD countries almost tripled over the 2000-04 period, reaching 43 GW in 2004. During the period 1990-2004, wind power capacity increased at an average annual growth rate of 23%, the fastest among all generating technologies. This phenomenal growth has largely been driven by government incentives, generally in the form of tax credits, production incentives or, more commonly, obligatory purchases supported by consumers through feed-in tariffs.

The capacity investments occurring in the last 15 years led to some significant changes in the OECD generation mix. The most noticeable trends are the increasing shares of natural gas-fired capacity and of renewables (especially wind), and the corresponding decline in the shares of nuclear, hydro, coal-fired and oil-fired capacities. In absolute terms, most generating technologies experienced increases in capacity. Tidal, wave and ocean based capacity increased only slightly, while oil-fired capacity actually declined.

The overriding question with respect to generation adequacy is whether OECD countries have the generation capacity necessary to meet current and future electricity demand. As is common practice, this report uses margins of installed capacity over peak-load as an approximate indicator of generation adequacy. Several countries currently enjoy comfortable reserve margins, and a few others experience rising reserve margins. However, in some countries, margins have trended downwards and are now at levels that could create risks for system reliability in the short term. These risks appear to be increasing with time, as projected demand outpaces projected capacity additions.

Reserve margins are good indicators to use in market monitoring and assessment. However, in competitive markets, a decline in reserve margins might not – and should not – be interpreted as a negative or undesirable market outcome. In some cases, it may represent market adjustments from an "overbuilt" position. In others, it may reflect a just-in-time investment approach intended to achieve better project economics and improved efficiency of the entire power system.

Additionally, reserve margins need to be assessed in the context of open electricity trade and increasing cross-border electricity flows. For example, total electricity exports from OECD Europe have increased by 3% per year during the 1990-2004 period. Anticipated moves towards further market integration will likely lead to even higher cross-border flows in many IEA jurisdictions. Open trade and cross-border reserve sharing allow for more efficient and integrated management of power systems.

Based on projected new capacity requirements, which reflect demand growth and projected plant retirements, it is clear that considerable funds will need to be invested in the electricity sector. The *World Energy Outlook 2006 (WEO 2006)* projects USD 11.2 trillion of investments in the global power sector to 2030, of which USD 5.2 trillion will be directed toward power generation (Figure 1.3). For IEA countries, the challenge is not so much the availability of funds. Rather, it is the ability to attract and reward investors in ways that adequately compensate for the risks associated with investments in power generation, when compared to those associated with alternative investments options.

Growth in electricity demand has been and remains the most important driver of investments in generating capacity. According to *WEO 2006*, OECD countries will need a total of 466 GW by 2015 to meet incremental demand: 47% in OECD North America, 41% in OECD Europe and 12% in OECD Pacific. For the period to 2030, total capacity requirements associated with increases in demand in the OECD will be 1 186 GW, with roughly comparable proportions on a geographical basis.

Figure 1.3

***Power generation accounts for almost half of projected global
power sector investment to 2030***

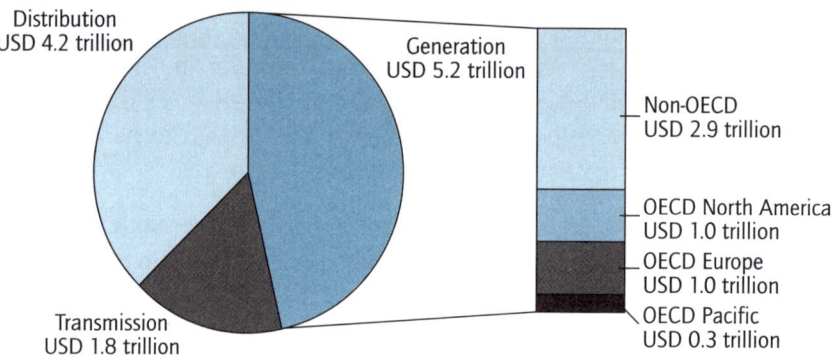

Source: IEA, 2006b.

Still, it is important to bear in mind that regional demand patterns can differ markedly. In spite of efficiency improvements, electricity demand has continued to grow in most IEA countries, although at generally lower rates than in the past two decades in most jurisdictions. In the last five years, North America experienced relatively faster growth than Europe, largely due to faster economic growth. Peak demand in some North American regions became more pronounced due to accelerated growth in cooling demand during the summer. Cooling demand is also becoming more significant in southern Europe.

The age of power generation units is one indicator of the need for potential decommissioning, which has significant implications for new capacity requirements. According to a Platts database (Platts, 2005), some 638 GW of installed coal, oil, gas and nuclear capacity in the OECD is now more than 30 years old. This represents about 27% of total installed capacity. The stock of coal- and oil-fired capacity is of particular concern; roughly half of the installed capacity is already more than 30 years old. In *WEO 2006,* projections of plant retirements are based on the assumption that fossil-fuelled and nuclear plants will be automatically retired at the end of their useful life, assumed to be 45 years. These projections are relatively uncertain. In reality, plant retirements are business decisions and factors such as stricter environmental regulation can lead to early plant retirements. Thus, some power plants may be retired earlier than planned while others will remain in operation beyond their projected useful life but will then require investments in refurbishments. For OECD countries as a whole, projected plant retirements over the period to 2015 amount to 215 GW or 9% of total capacity (in 2004). About 50% of these retirements will occur in OECD Europe, 42% in OECD North America and 8% in OECD Pacific. Looking forward to 2030, a total of 872 GW will need to be retired in similar proportions across the OECD regions.

Considering the needs to meet both incremental demand and plant retirements, *WEO 2006* projects that the OECD as a whole will require 681 GW of new capacity by 2015 (2 058 GW by 2030) of which 45% will be in OECD North America, 44% in OECD Europe and 11% in OECD Pacific (similar shares by 2030).

Key Message

Growing demand for energy, coupled with the need to replace or refurbish ageing infrastructure, creates significant investment requirements in power generation. Net capacity additions are not keeping pace with growth in demand and reserve margins are declining in some regional markets. If current market trends continue, there are risks of under-investment resulting in reserve margins falling to dangerously low levels.

In recent years, investments in power generation have not been widely diversified. Most of the investments in OECD countries, particularly in the last decade, have been in gas-fired capacity (mostly CCGT) although a shift is occurring in favour of coal-fired generation. In renewables, most of the investments have been in wind power. Considering the long lead times required for siting, approvals and construction of power plant projects, governments must act now to ensure that timely investments occur at the most efficient locations, and directed towards fuels and technologies that

OECD Europe at Multiple Crossroads.....................................

The power sectors in OECD countries operate under multiple policy objectives. The impact these policies have on power generation investment is currently perhaps most evident in OECD Europe. The European Union (EU) and its member states have implemented a series of directives and policies that aim to enhance efficiency, reliability and environmental sustainability.

A comprehensive EU energy policy package, proposed by the European Commission (EC) in January 2007, reinforces these policies and objectives. The most prominent directives and policies are:

- EU Market Directives drives liberalisation of EU electricity and gas markets.

- The EU Emission Reduction Allowances Trading Directive establishes the framework for the European Union Emission Trading Scheme (EU ETS).

- The Renewables Directive sets tentative renewables targets for member countries. A binding target was decided in the new policy package.

- The Large Combustion Plant directive sets standards for air pollutants for plants above 50 MW.

- EU green papers and policies on energy efficiency aim for 20% energy savings by 2020, compared to a base case. This corresponds to almost stable electricity demand towards 2020[2].

- National policies call for nuclear phase-out in some countries and for revisiting nuclear power in others.

These Europe-wide policies have had significant impact on power generation investments in Europe over the last 10 years and are expected to have greater influence in the coming decade. Some of the policies have been highly effective and have had precisely the intended impact. However, they also create a degree of uncertainty that carries important indirect impacts. In particular, uncertainty regarding new investments arises from lacking or inadequate implementation of market directives, national differences in the implementation of the EU ETS, lack of clarity regarding renewable and efficiency targets, and the ongoing debate about the future of nuclear power.

EU electricity market liberalisation, starting with the first EU Market Directive in 1996, was probably the single most important policy driver. The liberalisation process was intensified with the second directive in 2003. The EC conducts annual benchmarking reports on the progress of implementation of the market directives and the development of markets in the spirit of the directives. The first six reports published demonstrate significant improvements. However, electricity markets are deemed satisfactorily open and competitive only in

---

2. According to the European Commission Energy Efficiency Action Plan, transport and electricity consumption are the prime targets for improvement. Transport represents 20% of total energy demand in the EU. The saving potential in transport contributes 27% of total savings below the base case scenario. All in all, the goal of saving 20% of total energy implies a 15-20% reduction in electricity demand below the base case scenario. Again, this implies almost stable electricity demand from 2006 to 2020.

a few countries. In 2006, the EC's General Directorate for Competition (DG Comp) launched a sector inquiry with the aim of tracking behaviour and practices that undermine competition in European electricity and gas markets. Preliminary reports identify discriminatory foreclosure (effects of, for example, lack of effective unbundling and preferential long-term contracts) and lack of transparency as some of the most serious obstacles to competition.

Fully liberalised markets with effective market institutions aim to create the right conditions for promoting efficient investment. Starting from a state of overcapacity, which was the case in many European countries, a drive towards efficiency implies a period with no or few new-builds. Never the less, the large blackouts in 2003 and concerns for investment adequacy led to the development of the recent Security of Electricity Supply Directive. It adds some checks, balances and safeguards, but in general backs the main principles of market-driven investments as outlined in the market directive.

At the national level, several countries have, more or less explicitly, pursued objectives for the development of utilities to become so-called "national champions" – large companies that can take a significant role in domestic and international energy markets. Consolidation is a natural part of a healthy market development, and it is the role of competition authorities to ensure that competition does not suffer, resulting in a loss for consumers. However, nationally motivated interference in consolidation developments and in the work of competition authorities can undermine consumer welfare. Large consolidated companies may add value that can benefit consumers in the end, but when national champions are given value through beneficial national market positions, through, for example, a lack of unbundling and lack of third-party access to networks, it is at considerable cost for consumers. It is far from the spirit of the internal market, and the bill from compensation for stranded assets, poor competition and lack of market access is now fuelling scepticism of liberalisation among consumers in several countries.

Policies on climate change have also had an increasingly significant impact on the power generation sector. The EU ETS stands out as one of the more successful EU energy directives. Since 1 January 2005, when this sophisticated trading mechanism was successfully implemented on time, CO_2 emissions have carried a price. The price has oscillated between USD 1 and USD 40/tonne of CO_2, illustrating that market participants recognised CO_2 constraints in the system, as well as great uncertainty. The price of CO_2 emissions was passed through to wholesale electricity markets and, ultimately, to electricity consumers. This is the economically sound response and functioned exactly as intended. Some estimates indicate that the EU ETS has already reduced CO_2

emissions by some 100 million tonnes – about 2.5% of total OECD Europe emissions in 2004 (Walhain, 2006). The long-term effect should naturally spill over into investment decisions, providing incentives to invest in technologies that are less CO_2 intensive (*e.g.* renewables, nuclear, gas and high-efficiency coal), depending on the tightness of the CO_2 constraint.

Paradoxically, this successfully implemented tool risks having no, or perhaps even a negative, impact on investment. This fact is driven by three main concerns. First, the EU ETS is built on the Kyoto Protocol, which carries commitments only until 2012. This timeframe is not sufficiently long-term to support clean power generation investment. Second, the EU ETS is effective only if countries are willing to allow real CO_2 constraints. The system becomes distorted and the EU ETS loses its impact when some countries give all necessary allowances for free to new investments in, for example, coal plants. Third, CO_2 emissions were initially "grand-fathered" to emitters – emitters received emission allowances for free. With consumers now paying the full costs of CO_2 emissions, this adds considerably to the scepticism that arises from inadequate liberalisation.

The EU Renewables Directive, which came into force in October 2001, has also had a strong impact on investment. It proposes that member states adopt national targets for renewables that are consistent with reaching the overall EU target of deriving 21% of electricity from renewables by 2010. The EC assesses that, as long as current trends continue, member states will come close to the target. With the new energy package a binding target of 20% renewables in the energy mix by 2020 has been decided, but still not allocated to member countries. This target is based on the assumption that renewables could potentially provide about one-third of electricity generation in the EU.

The EU Directive on Large Combustion Plants defines a set of minimum requirements on emissions of air pollutants for existing and new combustion plants above 50 MW and will hence have a significant impact on the pace of plant retirements. All existing plants are obliged to comply with maximum emission ceilings for SO_2, NO_x and particulate matter by 2008. Alternatively, the plants must be part of a national emission reduction plan that may give non-complying units some scope to continue operating until 2015. Installation of flue-gas desulphurisation equipment for coal-fired plants is the most immediate and costly measure necessary to achieve compliance. In a later phase $deNO_x$ plants are required as well, and NO_x limits will become tighter from 2016.

Energy efficiency policies and targets on EU and national levels will have significant indirect impact on investments. The energy efficiency targets announced in the EC green paper imply almost unchanged electricity demand towards 2020; this can have an important impact in the short and medium term. Investors must be given a chance to assess the realism of such targets. Otherwise some of the benefits may be lost through inefficient investment responses.

National policies on nuclear power also have a significant impact on investment. Germany, Belgium and Sweden have decided to phase out nuclear power, yet that decision is constantly challenged on the national political scenes. In Sweden, several existing nuclear power plants have been allowed to undergo modernisation and upgrades, representing some of the most important investments in new generation capacity in the Nordic market. Spain has a moratorium on nuclear power but it is still not legally binding; the scope for modernisation, upgrades and life extensions is unclear. At the same time, a revival of nuclear power is being discussed in Spain. Prospects for new nuclear power are also under discussion in several other European countries, such as the United Kingdom, Turkey, Poland and the Netherlands.

This policy setting has already had important impacts on development of power generation capacity. Gas-fired combined cycles became the preferred technology, particularly in liberalising markets. Renewable policies drove their strong growth, particularly in wind power. Figure 1.4 shows the development of installed capacity in OECD Europe over the 1990-2004 period, according to technology type. Installed generation capacity increased by 143 GW over the last 15 years, from 641 GW in 1990 to 784 GW in 2004. About 88 GW came from increases in gas-fired capacity, of which 81 GW were CCGTs. Another 34 GW was from wind power. Other increases were from hydro (21 GW), nuclear (7 GW) and other renewables (11 GW). Oil and coal capacity decreased by 10 GW and 8 GW, respectively.

Germany has the largest generation park in Europe with significant shares of most generation types, most notably coal-fired generation. France has almost half of installed nuclear capacity in Europe. Italy and the United Kingdom have significant shares of installed gas-fired generation. In 2004, 143 GW of installed capacity in OECD Europe was gas-fired, of which 85 GW were CCGT. In 2004, the United Kingdom had 31% of installed CCGT capacity in OECD Europe, Italy had 23% and Turkey had 16%. CCGT capacity in Spain is increasing strongly. Oil-fired generation capacity represents 9% of total installed capacity and is relatively evenly spread across all countries, with significant shares in Italy, France and Spain. In most cases, it serves special purposes such as back-up and supply in small islanded systems. In 2005, it represented less than 4% of total electricity generation in OECD Europe.

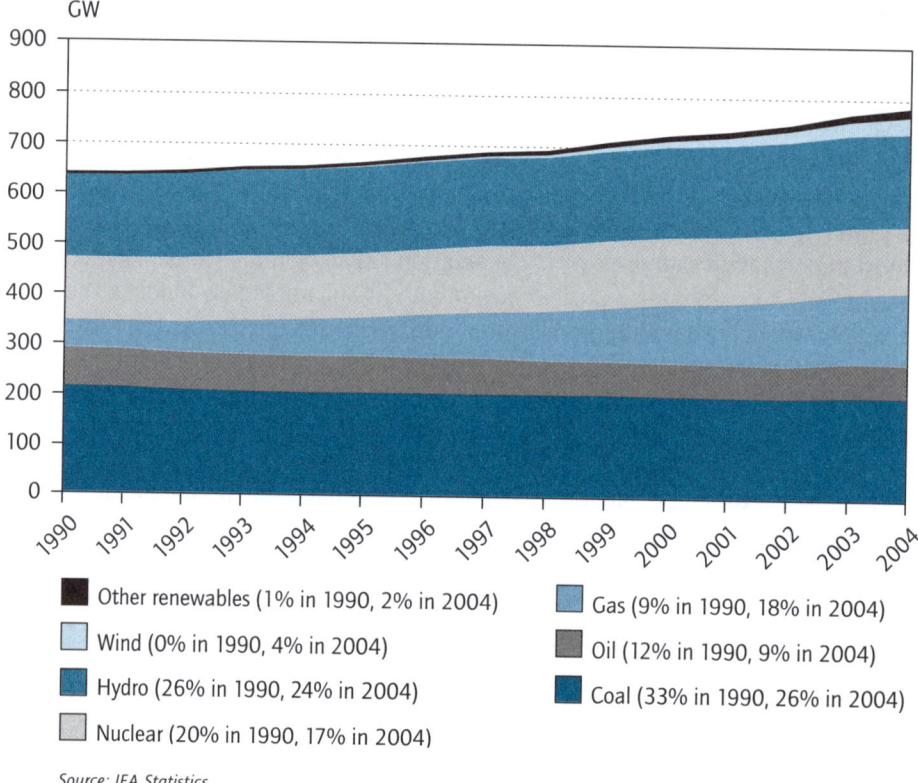

Figure 1.4

Gas and wind capacity increased, but coal maintains the largest share of installed generation capacity in OECD Europe

- Other renewables (1% in 1990, 2% in 2004)
- Wind (0% in 1990, 4% in 2004)
- Hydro (26% in 1990, 24% in 2004)
- Nuclear (20% in 1990, 17% in 2004)
- Gas (9% in 1990, 18% in 2004)
- Oil (12% in 1990, 9% in 2004)
- Coal (33% in 1990, 26% in 2004)

Source: IEA Statistics.

Figure 1.5 shows actual generation in OECD Europe since 1990 by fuel type. Germany and France generated about 1 200 TWh of power in 2005, representing about 34% of total generation in OECD Europe. France exported more than 60 TWh in 2004, approximately 2% of total generation in the region. Gas-fired generation is increasing at the fastest rate amongst the various types of generation, rising from 6% of total generation in 1990 to 20% in 2005. It is interesting to compare the shares of generation against the shares of installed capacity for various sources as this shows rates of capacity utilisation. Shares of gas and coal-fired generation more or less correspond with their shares in installed capacity. Nuclear power represents only 17% of installed capacity but 28% of generation, down from 30% in 1990. Hydro, wind and oil capacity represent far lower shares of generation than their corresponding shares of installed capacity.

Figure 1.5

Gas increases rapidly, while coal and nuclear maintain the largest shares of electricity generation in OECD Europe

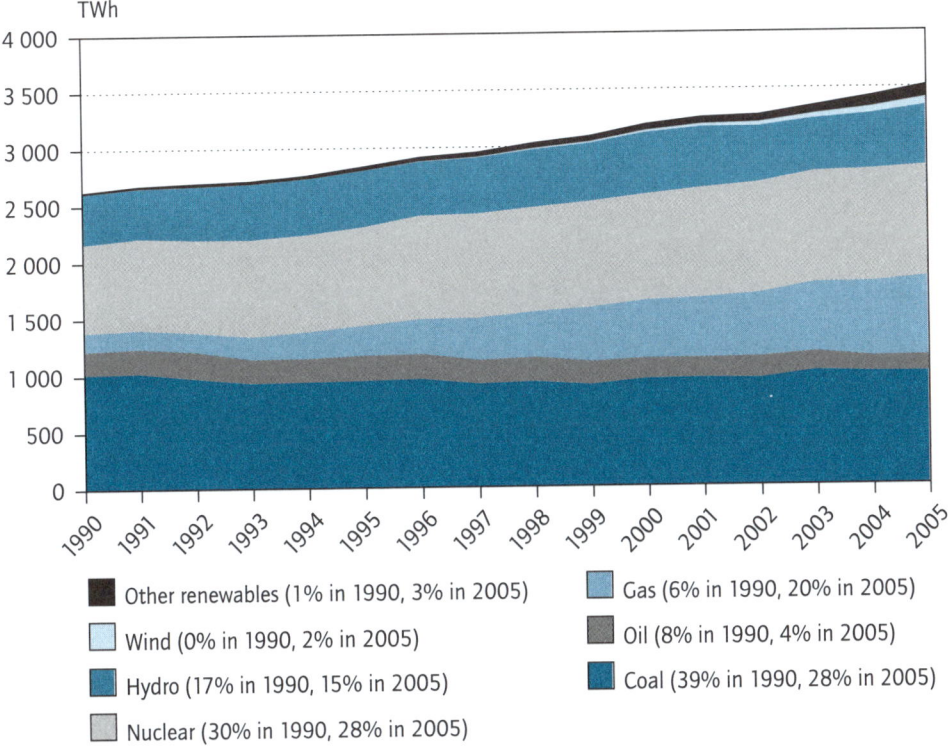

- ■ Other renewables (1% in 1990, 3% in 2005)
- □ Wind (0% in 1990, 2% in 2005)
- ■ Hydro (17% in 1990, 15% in 2005)
- □ Nuclear (30% in 1990, 28% in 2005)
- □ Gas (6% in 1990, 20% in 2005)
- ■ Oil (8% in 1990, 4% in 2005)
- ■ Coal (39% in 1990, 28% in 2005)

Source: IEA Statistics.

Overall, installed generation capacity in OECD Europe is well diversified. Coal-fired capacity dominated in both 1990 and 2004, although its share decreased from 34% to 27%. This diverse portfolio is also mirrored in several individual countries, but less so and with important exceptions. Almost 100% of Norway's capacity is hydro. Some 55% of installed capacity in France is nuclear, which accounts for almost 80% of the country's total generation. In the Netherlands, almost 70% of installed capacity is gas-fired. The generation mix in Europe is thus only well diversified if European countries are able to trade and co-operate, to spread risks and benefit from mutual advantages.

Some serious challenges ahead may constrain OECD Europe's ability to maintain the current level of diversification into the future. In the past 15 years, most new generation capacity was in gas and wind. Much of the generation capacity installed in the 1970s, a period marked by a major shift to coal and nuclear, is still operating. This capacity is now ageing and some is approaching the end of its useful life. Figure 1.6 illustrates generation capacity (coal, oil, gas and nuclear) that is now more than 30 years old as a share of total installed capacity.[3]

Figure 1.6

20% of coal, oil, gas and nuclear capacity in OECD Europe more than 30 years old

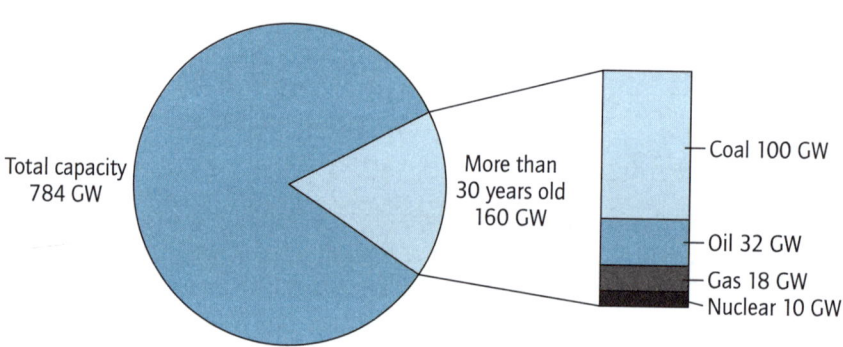

Sources: Platts, 2005; IEA Statistics.

About 160 GW of coal, oil, gas and nuclear capacity in Europe is more than 30 years old. This represents 20% of total installed capacity in 2004. More than half of European coal capacity is more than 30 years old; some 85% in the United Kingdom, some 50% in Poland, and some 40% in Germany.

Coal-fired generation is often technically able to operate for as long as 60 years, and longer with investments in refurbishment. But the performance gap between old and modern units increases both from wear-and-tear of existing plants and from advancements in new technology. The gap in environmental performance is even starker. The higher efficiency of modern units reduces the greenhouse gas footprint, and most old units lack modern cleaning capabilities such as flue gas desulphurisation equipment. Owners of older coal-fired units will need to carefully consider the economic viability of investing considerable sums in old, often small, and relatively inefficient units to meet minimum environmental requirements. Without a doubt, some will close rather than making the costly upgrades. For example, in the United

---

3. About 110 GW of installed hydro capacity is also more than 30 years old. However, the age of hydro plants is less relevant as an indicator of a need for replacements, even if investments for refurbishments and upgrades of hydro capacity are also important and necessary.

Kingdom, it is expected that 8 GW of coal capacity, about one-quarter of total UK coal capacity, will close by 2015 (IEA, 2007a).

More than 90% of nuclear capacity in OECD Europe is still less than 30 years old, but 60% is older than 20 years. Much of this capacity was built during the late 1970s and early 1980s, and is now approaching a phase of needed investments for refurbishment, upgrading and life time extension. Some of the oldest units will be closed down. In countries with moratoriums and decisions to phase out nuclear power, investment in refurbishment may not take place. Under the current phase-out agreement in Germany about 12 GW of nuclear power will be decommissioned by 2016 effectively deferring some otherwise economic investments for refurbishments. In the United Kingdom, about two-thirds of the installed nuclear capacity (about 8 GW) is expected to be decommissioned by 2014, and most of the remaining capacity within a decade thereafter.

The IEA's *World Energy Outlook* examines energy trends under current policies and compares them to trends that could be realised if more policies under consideration by governments were implemented. The publication refers to these two projected outcomes as the "reference" and "alternative" scenarios, respectively. According to reference scenario projections in *WEO 2006*, 294 GW of new capacity will need to be built in OECD Europe by 2015: 106 GW to replace decommissioned plants and 188 GW to meet increasing demand. By 2030, OECD Europe will need to build 928 GW, much of it to replace the 435 GW projected to be decommissioned, which is more than half of currently installed capacity. The *WEO 2006* pointedly emphasises that the existing policies on which the reference scenario projections are based reflect "an expensive, dirty and unsecure path." In the *WEO 2006* alternative policy scenario, one of the most important groups of alternative policies taken into account is those directed towards improving energy efficiency. Decisive measures to increase energy efficiency could significantly reduce the need for new capacity: to 225 GW by 2015 and to 713 GW by 2030. The energy efficiency targets expressed in the recent EC Green Paper on Energy Efficiency propose a potential for even larger and faster improvements in energy efficiency.

It is clear that the investment requirements over the next decade are substantial. Within a liberalised market framework the investment incentives come from the price of electricity. Wholesale electricity prices in most European countries started to increase in 2004; they have since been volatile and remained at a substantially higher average level. The price increase can be directly attributed to three underlying factors. Gas and coal prices started to increase significantly from 2004, increasing the cost of generating electricity. The 2005 launch of the EU ETS effectively internalised a CO_2 price in the wholesale electricity price, thereby pushing the price higher. Finally, tighter balance between supply and demand in European electricity markets included more expensive resources in the electricity dispatch stack. These price increases prompted the EC to launch a sector inquiry to investigate the presumption that lack of competition was also an explanatory factor. Under effective market conditions, the market should respond to these signals by adding new capacity and rebalancing the preferred generation options – in this case, to reduce dependence on gas-fired generation and favour generation with a low CO_2 footprint. Indeed, market players in some parts of Europe have responded and in other parts they are starting to propose new projects.

According to the Platts database (Platts, 2005), more than 35 GW of new capacity (coal, oil, gas, nuclear and hydro) was under construction in OECD Europe in 2005. More than half of this was gas-fired, and most of these were CCGTs (some 20 GW). Substantial wind power capacity is added every year. In 2006 alone, 7.5 GW were added in the EU as follows: 2.2 GW in Germany; 1.5 GW in Spain; and between 1.0 GW and 0.4 GW each in France, Portugal, Italy and the United Kingdom (EWEA, 2007). The first evidence of a shift in focus from gas to other sources is materialising. In 2005, Finland took the decision to construct a new nuclear reactor – the first such decision in OECD Europe in more than 10 years. France is in the late-stage planning process for another nuclear reactor; the final construction decision is expected in 2007.

The shifting focus of fuel and technology choice is more evident when looking at planned plants. Coal is again becoming a preferred technology compared to the previous almost exclusive concentration on new gas-fired generation. According to the Platts database, some 120 GW of new capacity was planned in 2005 although not all of this can be expected to come to fruition. Most of the confirmed plants are due to be commissioned in the 2010-15 time-frame. Of the 120 GW planned capacity, 80 GW was CCGTs and 22 GW was coal-fired. A high share of the new investment activity is proposed in Germany. According to the German Electricity Association (VDEW), investors have announced that 31.4 GW of new generation capacity will be commissioned by 2012, of which roughly one-half is coal-fired, one-quarter is gas-fired and the rest is renewables, mainly wind. Another 13 GW by 2016 (VDEW, 2006). Several thousand GW of wind power are planned, including large offshore farms mainly off the costs of northern Germany, Denmark, Sweden, the Netherlands and the United Kingdom. The offshore farms have been in the planning phase for a long time, but so far only a few large farms in Denmark and a few smaller farms in the United Kingdom have materialised.

Investors in several countries, including the United Kingdom, Germany, Norway and Denmark, are also moving forward with investment decisions and plans for carbon capture and storage (CCS). Projects will only be for demonstration by 2015; however, they may start to influence the prospects for clean investments shortly thereafter. Increased certainty regarding CO_2 constraints beyond 2012 (*i.e.* beyond the Kyoto Protocol) will be critical for these decisions. The EC's new energy policy package includes a proposal to stimulate the construction of 12 large-scale CCS demonstration projects. This is to be followed up with a clear out-look for future CCS policies with the aim of being able to set requirements that all new coal-fired plants be fitted with CCS by 2020.

Changes in demand patterns also played a role in the development of installed capacity and actual generation output. From 1990-2004, electricity demand in OECD Europe increased by 1.8% annually. Peak-loads aggregated across all individual countries increased by 1.7% – slightly less than total demand, indicating lower peak demand in aggregate. However, demand patterns vary considerably from country to country. For example, over the past five years in winter-peaking France, annual peak demand increased at a rate 1% higher than the increase in total demand. Annual peak demand in Portugal, (also winter-peaking) increased almost 2% faster than total annual demand during the last five years. Peak demand is also increasing faster than annual electricity demand in some summer-peaking countries. In Greece, annual peak demand (summer) increased 0.7% faster than total demand over the last 15 years. Increased use of air conditioning is one factor driving increases in summer peak demand. In fact, Greece, Italy and Spain have switched and Turkey is close to switching from winter- to summer-peaking.

Demand profiles are very diverse across OECD Europe. Demand peaks and distribution of the load curve throughout the day, season and year depend on numerous factors that vary widely across Europe. Temperatures are different and have different effects. Some countries have high shares of industrial consumption while others have much electrical heating or electrical cooling. Meeting peak-load on a national basis may put high demands on national electricity systems, but many of the challenges can be reduced considerably in aggregate across several countries. The co-operation in the Union for the Co-ordination of Transmission of Electricity (UCTE)[4] representing systems in the synchronously interconnected network in continental Europe is a good

---

4. UCTE members include system operating companies in Austria, Bosnia Herzegovina, Belgium, Bulgaria, Switzerland, the Czech Republic, Germany, Denmark West, Spain, France, Greece, Croatia, Hungary, Italy, Luxembourg, Montenegro, Former Yugoslav Republic Of Macedonia, The Netherlands, Poland, Portugal, Romania, Serbia, Slovenia and the Slovak Republic.

example. Figure 1.7 shows load in UCTE during hours with highest demand in every month of 2006. It shows the highest demand both for the aggregate of individual hours and the highest demand for UCTE in total.

Figure 1.7

Important gains from UCTE co-operation evident in lower aggregate peak-loads compared to individual

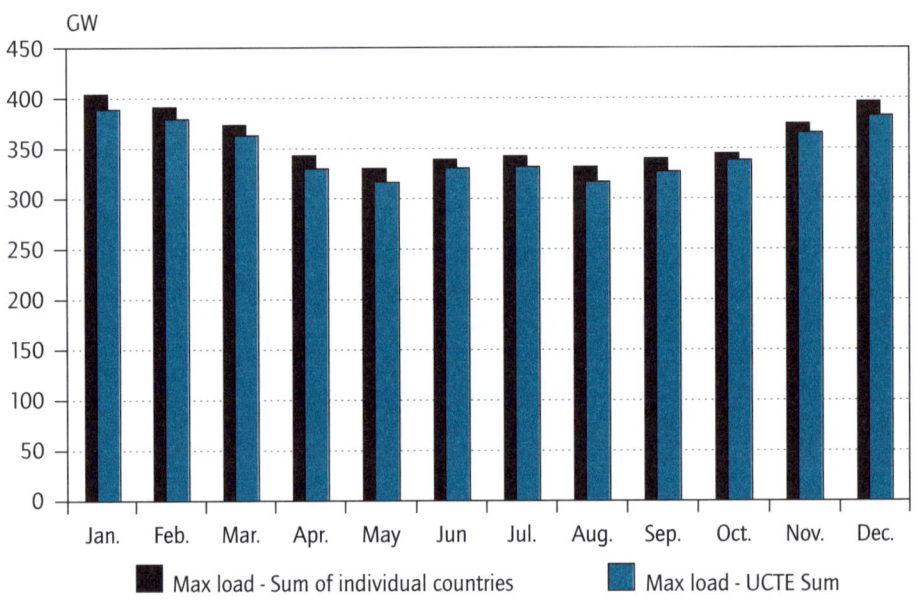

Max load - Sum of individual countries   Max load - UCTE Sum

Source: UCTE.

UCTE in aggregate is winter peaking. The relative peak-load of individual countries is reduced when placed in the context of a larger area. In 2006, the sum of individual peak-loads for each UCTE member was 413 GW. However, coincident peak-load – the actual peak that occurred in the whole system – was 391 GW, 22 GW (5%) lower. The difference between peak-load in individual countries and peak-load across the much larger UCTE system underlines the potential benefits from co-operation, trade and sharing of reserves. At present, availability of transmission interconnectors constrains the potential to fully benefit from such sharing of reserves.

Figure 1.8 compares the development of installed generation capacity in OECD Europe to the development of aggregate peak-load. It shows total installed capacity and the margin of installed generation capacity over aggregate peak-load. The reserve margin is a commonly used, but crude indicator of generation adequacy. Several factors make it difficult to interpret. Cross-border trade in interconnected systems makes it more relevant to analyse peak-load on a system-wide basis, rather than aggregate peak-load but such historic data is not currently available. In addition, installed capacity is not of much use to meet demand if it is, in effect, not available. Availability of installed capacity depends on several factors. For example, one feature of wind power is that it is available more or less by chance. Figure 1.8 also includes the development of the margin of generation capacity over aggregate peak-load assuming that wind power is available at only 20% of its installed capacity. This roughly corresponds to the average availability of installed wind power in Europe; in fact, a lower availability must be expected at peak-load.

Figure 1.8

Capacity margins in OECD Europe fall
when low availability of wind power is taken into account

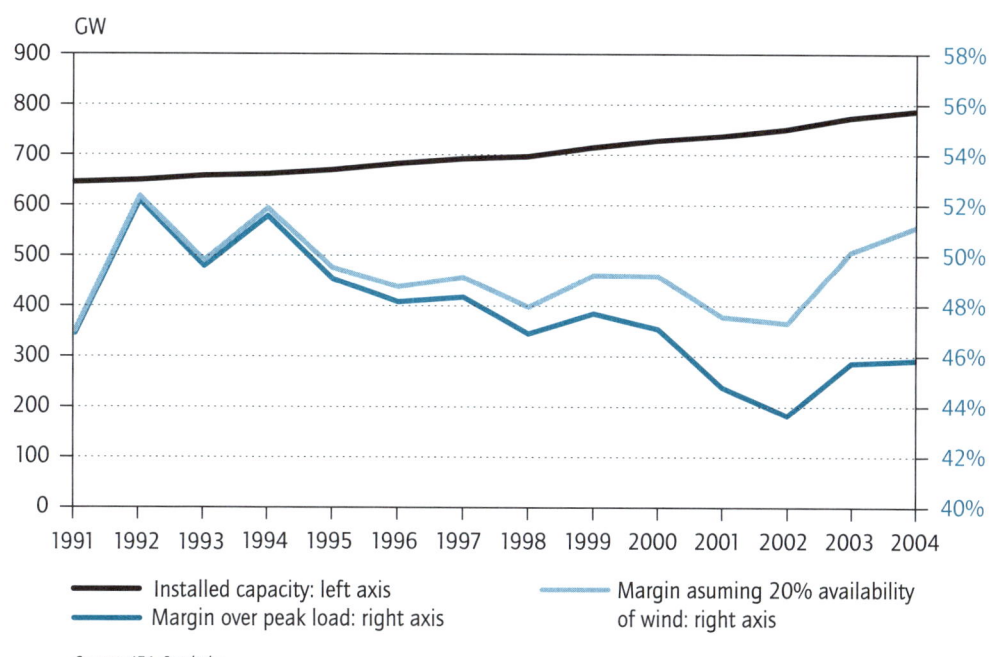

Source: IEA Statistics.

The relatively constant margin of installed capacity over peak-load from 1991 to 2004 shows that development of installed capacity has kept pace with the increase in peak-load. However, much of the new installed capacity comes from wind power. If the particularly low availability of wind power to meet peak demand is taken into account, it is clear that reserve margins have decreased. Decreasing reserve margins are certainly to be expected when introducing competition, and are not necessarily a concern. Several OECD European countries had very high capacity margins prior to liberalisation, when incentives to minimise investments and costs were weaker. In addition, cross-border trade and co-operation enable reserve sharing and dispatch optimisation across a larger area. However, at some point new investments are needed to maintain a reliable electricity system. In OECD Europe, this point is approaching.

Every year, the association of European Transmission System Operators (ETSO) publishes a report on generation adequacy, based on inputs from the various member regions: UCTE, Nordel (the Nordic countries), the Baltic States, Great Britain and Ireland. As yet, there is no common methodology to assess generation adequacy across all ETSO member systems. The UCTE methodology is used as the reference. The key measure, "reliably available generation capacity", is assessed by deducting unavailable capacity from installed capacity (unavailable capacity includes wind plants not expected to be available, mothballed plants, capacity out for refurbishment, assessment of outages and capacity reserved for ancillary services). "Remaining capacity" is calculated as the margin of reliably available generation capacity over peak-load, plus a safety margin.

Projections show a positive remaining capacity across the ETSO area until 2012. By 2015, it will be necessary to commission 20 GW of new generation capacity, in addition to plants already under construction. A more optimistic scenario for remaining capacity is calculated by including planned plants that are reasonably certain (i.e. plants that will be built to fulfil national renewables targets and projects that have requested connection to the transmission grid). This scenario shows a positive margin of almost 20 GW of remaining capacity in 2015 (ETSO, 2006). This confirms the impression that the pool of planned plants is large enough to meet demand. However, important final decisions on additional investment must be made in the next few years to ensure that a sufficiently large share of planned plants actually materialise.

This reassuring picture has two critical caveats. One is that system operators do not have much reliable information about future decommissioning of plants. Thus, the EU's Large Combustion Plant Directive will have an impact on business decisions to maintain or decommission existing large plants. Other plants are also likely to be decommissioned, although it is difficult to accurately predict the exact capacity and timing. The other important caveat is the considerable regional and national differences.

Starting from north, the Nordic countries in Nordel do not foresee adequacy problems before 2015. Installed generation capacity and capacity under construction are expected to be sufficient to meet peak demand. Another 700 MW planned capacity expected to come to fruition adds even more generation margin over peak-load. Despite the comfortable reserve margin, due to the high share of electrical heating, the Nordel system may need to rely on imports during extreme weather conditions, such as a one-in-ten year cold snap.

The United Kingdom will need additional final investment decisions to meet peak demand as soon as 2008. The shortage point is shifted to 2009 when probable plants are taken into account. With such short time horizons, some new investment projects will need to be CCGT, as was the case in the past. Taking the 2 GW import capacity from France into account provides some additional leeway, but the expected decommissioning of 16 GW by 2015 puts considerable pressure on the need for new investments in the United Kingdom in the medium term. Investors have responded adequately and just-in-time in the recent past.

Ireland already relies on imports from Northern Ireland during peak periods. The situation is manageable to 2010, but investment decisions must be taken soon to avoid increasing the risk for real shortages to unacceptable levels.

The main UCTE block (Austria, Belgium, Bosnia-Herzegovina, Croatia, France, Germany, Luxembourg, the Netherlands and Switzerland) will be able to meet peak-load during the 2006-10 timeframe, but will be 11 GW short by 2015 if probable plants are not taken into account. Today, this main block is a net exporter. Without additional investment decisions this capability will be reduced and perhaps even reversed by 2015.

On the Iberian Peninsula peak-load is met until 2010, after which new investment is needed. By 2015, 12 GW of new capacity is required to meet peak-load, which will have shifted to a total Iberian summer peak.

Italy has already commissioned or started construction on a considerable amount of new investment projects. These reflect an investment boom after the 2003 black-out. Italy will meet peak demand by 2015 with existing plants and plants under construction.

South eastern UCTE (the Former Yugoslav Republic of Macedonia, Greece, Montenegro and Serbia) already relies on imports to meet peak demand. In 2006, there was a shortage of 3 GW to meet summer peak demand. New investment decisions have been made in Greece but further investments are urgently required.

The CENTREL block (Poland, Hungary, Slovakia, Czech Republic and Western Ukraine) has plenty of margin to meet peak demand, which will remain positive beyond 2015.

In Turkey, demand increased rapidly over the last decade with a short interruption during the economic recession in 2000/2001. Investments are necessary to keep pace with increasing demand. At the same time, there is considerable scope to increase efficiencies and utilisation of existing coal-fired plants.

Overall, generation margins are falling in OECD Europe, according to ETSO's 2006 generation adequacy report, with the tightest balances in Ireland, Greece, Spain and the United Kingdom. New generation capacity must be added by 2015. The capacity gap reported by ETSO amounts to 20 GW, which is relatively modest given the timeframes available for response. However, some major challenges are likely to arise just after 2015, at which point decommissioning of old coal plants and phase out of nuclear power are likely to accelerate. Transmission systems must evolve in tandem with the development of generation capacity, particularly considering that strong growth in wind power is likely to continue.

There is scope for improving the use of cross-border transmission capacity to share resources and to optimise plant dispatch across larger areas in most regions in OECD Europe, particularly within the UCTE-synchronised grid. Significant scope also exists for improving trade within the day of operation and in real-time. At present, cross-border trade for real-time balancing is almost non-existent in most parts of Europe. Improved cross-border trade would minimise investment costs and improve system security; however, it requires even stronger co-operation and co-ordination among system operators. Lack of co-ordination transforms cross-border trade into a weakness that jeopardises system security. This point is well illustrated by a large disturbance in the UCTE system in November 2006, which disconnected 15 million consumers from the grid for one to two hours. European regulators attributed the disturbance to a failure to adequately co-ordinate system operations (ERGEG, 2006b).

With the projected strong increase in wind power, presumably concentrated in areas with large wind resources, improved cross-border trade and co-ordination is not an option. It is a requirement. Thus, it is imperative to provide appropriate and undistorted incentives for wind developers, system operators and transmission developers. The most important source of such incentives is effective trading mechanisms that provide clear, transparent and dynamic price signals, including strong locational signals.

OECD North America Shifts Focus Away From Natural Gas

Over the last decade, North America has seen an unprecedented build-up in gas-fired capacity. Throughout the United States and Canada, capacity additions have been predominantly in CCGTs. Growth in wind power has also been phenomenal, especially in the 2001-06 period. Despite significant capacity additions, reserve margins in many jurisdictions have declined as peak demand has continued to grow.

The next ten years will likely see a shift away from natural gas in the United States, as the US generation sector is turning its interest towards building new, more efficient coal-fired power plants and considering building new nuclear reactors. The trend towards more renewables, especially wind power and biomass, is expected to continue across North America. In Canada, further development of gas-fired capacity will likely be accompanied by potential hydro projects in Manitoba, Quebec, Newfoundland and Labrador. Although the US and Canada differ markedly in terms of commitment to the Kyoto protocol, both countries are actively pursuing efforts to promote investments in clean generating technologies, including renewables, carbon capture and storage, and nuclear.

Canada

Canada has seen considerable development of new generating capacity in most provinces, mainly natural gas, renewables (mostly hydro and wind power) and nuclear (through refurbishments). Capacity additions amounted to 10 GW during the 2000-05 period. Investments in capacity have been particularly strong in Quebec (the largest electricity market in Canada), Ontario (where demand has shifted towards summer peak) and Alberta (with highest provincial growth in electricity demand). Wind power investments accelerated in the 2004-06 period in most provinces including the Maritimes, with Quebec leading in total installed capacity.

Figure 1.9

**Projected generation capacity in Canada to 2015
shows further development of renewables and gas-fired capacity**

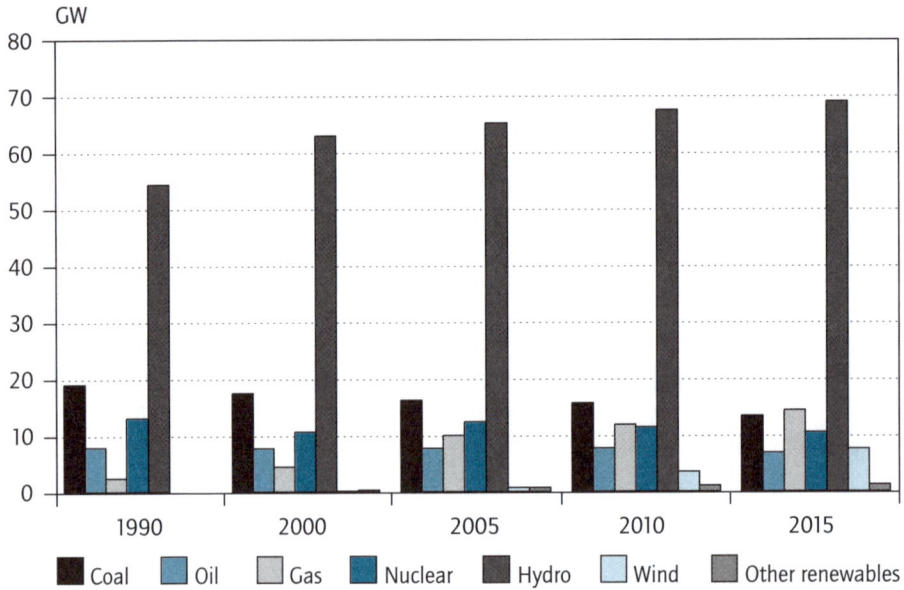

Source: Natural Resources Canada (2006).

Canadian electricity markets have developed along provincial boundaries; the extent of power sector reform towards market liberalisation has varied across the country. Ontario and Alberta have moved furthest along the reform path. Alberta has established fully competitive electricity markets at both the wholesale and retail levels. Legislative changes have transformed Ontario's electricity market into a hybrid competitive/regulated market. Other provinces have chosen to partially restructure their markets or to maintain the status quo. British Columbia, Saskatchewan, Quebec and New Brunswick have wholesale access and retail access to industrial customers; Manitoba allows only wholesale access. Most provinces continue to consider structural adjustments and improve market design rules.

As is the case in other IEA countries, increases in generating capacity in Canada are needed to meet growing demand and to compensate for anticipated plant retirements. Overall, according to the Platts database (Platts, 2005), 54% of coal-fired power units are more than 30 years old, compared to 41% for oil and 17% for gas. Most nuclear reactors are less than 30 years old. In Ontario, the supply challenge is further compounded by a provincial government decision to phase out 6 300 MW of coal-fired units by 2014 as a means of cutting CO_2 emissions that originate from the power sector. This capacity replacement need creates significant investment opportunities for less carbon-intensive generation technologies in the province.

Hydroelectric production is concentrated in British Columbia, Manitoba, Quebec, Newfoundland and Labrador. Generation in Alberta, Saskatchewan, Ontario and the Maritimes is largely thermal based. Nuclear generation plays an important role in Ontario and New Brunswick and a limited role in Quebec. Alberta may see emergence of nuclear generation to meet the energy needs for oil sands development. Natural gas-fired generation and wind power have trended upwards in most provinces. Future investments in British Columbia will be influenced by the new provincial *Energy Plan: A Vision for Clean Energy Leadership* (issued in February 2007), that requires that all new electricity projects developed in the province will have zero net greenhouse gas emissions, while clean or renewable electricity generation will continue to account for at least 90% of total generation.

According to the North American Electric Reliability Council (NERC), peak demand in Canada is expected to increase by 13% or 9.5 GW during the period 2006-15. Over the same timeframe, planned capacity additions will increase by about 7 GW, of which about 90% are natural gas-fired and wind-powered facilities. The plants currently under construction or planned will not adequately cover demand growth and decommissioning, suggesting a need for additional capacity by 2015. Assuming no additional capacity is commissioned, NERC estimates that available capacity margins will remain steady until 2009 in the range of 12-14% and decline in the longer term to just above 10% by 2015 (Figure 1.10).

Figure 1.11 provides the projected peak demand in Ontario to 2012, under normal and extreme weather conditions, the latter of which can have significant impact on demand. The independent electricity system operator of Ontario (Ontario IESO) projects that summer peak demand in the province will increase by 1.1% annually and winter peak demand by 0.7% annually over the period to 2015.

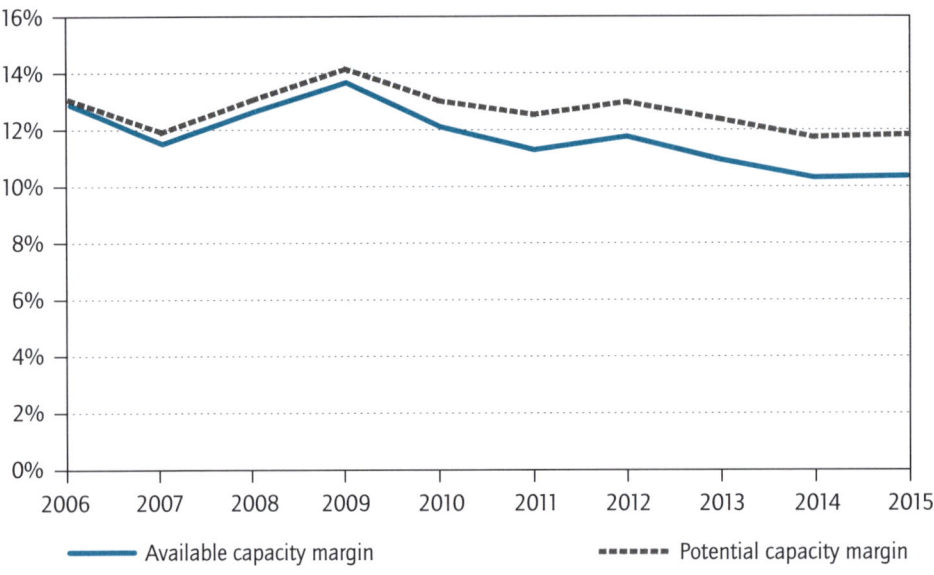

Figure 1.10

Reserve margins steady to 2010 and declining in the longer term in Canada

Source: NERC, 2006.

United States

During the last 10 years, the United States has seen an unprecedented level of investments in gas-fired generation capacity, facilitated by an integrated North American gas pipeline network that allows for substantial gas imports from Canada. Market reform also drove this "dash for gas" as the fuel of choice for the majority of new generating units, particularly in that it paved the way for independent power producers (IPPs) to compete with incumbent utilities. As a result, gas-fired generating capacity has increased by approximately 80% since 1999. The construction boom in gas-fired power plants between 1995 and 2001 led to an overall surplus of generating capacity in the US electricity industry in 2004/2005. The gas share in the generation mix increased from 14.2% in 1994 to 18.7% in 2005 and the gas share of total installed capacity increased from 21% to 40% over the same period.

States in the Northeast (*e.g.* New England, New York, and those in the PJM Interconnection – such as Pennsylvania, New Jersey and Maryland) and the Midwest, as well as Texas, have been most active in implementing power sector reforms and developing competitive markets. As a result, competitive

Figure 1.11

Projected electricity demand in Ontario is increasingly susceptible to peaks and vulnerable to extreme summer heat

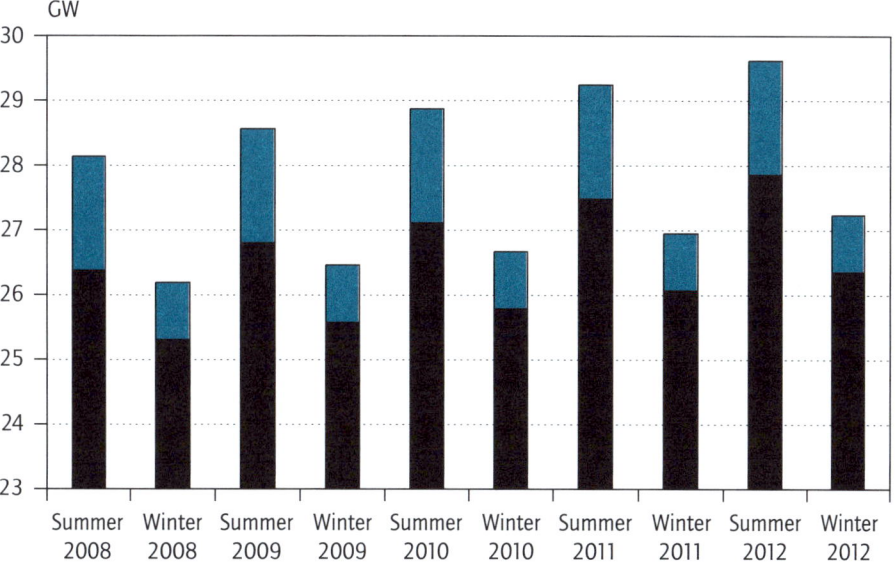

Source: Ontario IESO.

markets are now firmly established in these states. However, some other states (*e.g.* California, Virginia and Montana) continue to question the benefits of electricity market reforms, particularly in light of concerns about surging electricity prices and the fall-out of the Enron crisis.

In the United States, as in many other IEA countries, investments in power generation have significantly outpaced those in transmission. In conjunction with rapid growth in demand, this has created transmission congestion problems in a number of urban areas (*e.g.* New York City, San Francisco, Boston and New Orleans) and across broader regions (*e.g.* Southern California and Southwest Connecticut). In August 2006, the US Department of Energy released a study on transmission congestion and is now developing approaches to alleviate it, including the designation of "national corridors" (areas with critical congestion problem).

According to the Platts database (Platts, 2005), the age distribution of plant facilities in the US is as follows: 57% of coal-fired units are more than 30 years old, compared to 72% for oil and 24% for gas. About 30% of nuclear reactors are more than 30 years old. The age of the coal fleet and the business decisions to decommission old units must be considered within the context of the implementation of the *Clean Air Act* (CAA). The CAA mandates best existing technologies for air pollutant abatement on all new and significantly modified generation units. Initially, the definition of "significant modification" effectively made it possible to avoid installing the latest air pollutant abatement technology on existing units built before 1977. In the late 1990s, litigation against several utilities provided the impetus to launch a large-scale enforcement initiative and force a more narrow interpretation of the CAA. To date, the US Environment Protection Agency has settled with 11 utilities for an estimated USD 5.6 billion, all of which is to be invested in air pollution control. Some cases are still pending but a US Supreme Court ruling during 2007 is likely to further reduce uncertainty about the interpretation (Wilson and Potts, 2007).

Renewed interest in nuclear power is attributed to several factors including: concerns over energy security, partially triggered by the sharp increases in fossil fuel prices, and rising CO_2 emissions. Despite a lack of investments in new nuclear generation capacity in the last two decades, the United States remains the top nuclear generator in the world. In 2005, it had 98.3 GW of capacity that produced 809 TWh of power, providing about one-fifth of US electricity supply. In recent years, significant effort has been undertaken to improve capacity utilisation factors and upgrade nuclear plants, thereby allowing nuclear generation to maintain its 20% share from 1994 to 2005, despite no new plants and rising demand.

The *Energy Policy Act* (2005) includes specific incentives for investments in new nuclear power plants. It extends the Price-Anderson Act for a period of 20 years, which limits liability to third parties to about USD 10 billion. The act also offers a production tax credit of USD 0.018/kWh for the first 6 000 MW of new nuclear plants during their first eight years of operation, as well as federal risk insurance totaling USD 2 billion to cover regulatory delays of the first six advanced nuclear plants. In addition, the act provides federal loan guarantees for up to 80% of the project cost of advanced nuclear reactors (or other emission-free technologies).

In 2005, capacity additions were 17 622 MW of which 84% was gas, mostly CCGT units. The total figure includes 2 000 MW of wind power (mainly in Texas and Oklahoma) and 415 MW of coal-fired capacity. Plant retirements amounted to 3 172 MW. In 2005, total generation was up by 2.1%, almost identical to an annual average of 2% in 1994-2005. The most significant increase was in wind power, at 11.6% in 2005 and 23.9% in 2004. Wind power growth in recent years is driven by incentives provided by the renewable portfolio standards in many states and through a federal production tax credit, which was extended to 31 December 2008. However, it remains a small share of total generation. Generation by coal-fired plants was 19% higher in 2005 than in 1994. Capacity utilisation of coal-fired plants increased from 62% in 1994 to 73% in 2005.

Figure 1.12

US capacity additions to 2015:
Focus shifting towards coal after recent boom in gas investment

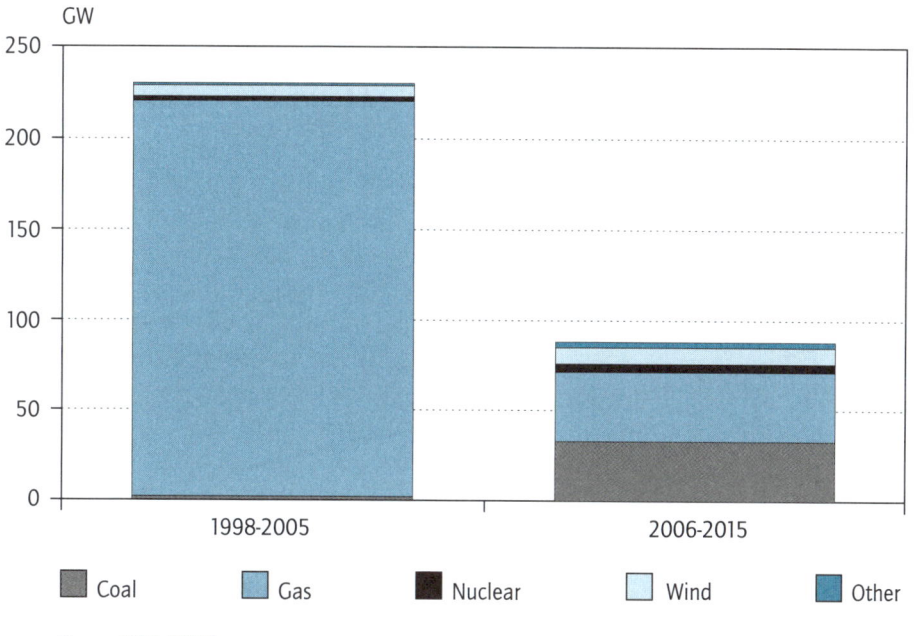

Source: NERC (2006)

According to NERC, planned capacity additions in the United States during the period 2006-15 will total 88 107 MW, with gas (mostly CCGT) showing a slight lead over coal (Figure 1.12). Recent planned investments have shifted more towards coal-fired power plants. The projection includes 12 000 MW of renewable power (predominantly wind and biomass). Most of the proposed capacity is scheduled to start commercial operation by 2012. The new coal-fired power plants, which represent more than one-half of all proposed capacity additions, will be concentrated in Texas, Illinois and Kentucky. According to the World Nuclear Association, as of January 2007, there were 23 nuclear reactors planned and proposed in the US, for a total capacity of 26 716 MW. NERC projects that some 5 GW of nuclear capacity will be added during 2006-15 corresponding to slightly less than the first six nuclear reactors that will receive special support from the new Energy Policy Act. More than half of the new reactor sites currently being considered by energy companies are located in the Southeast. Four entities have submitted site review applications for at least five nuclear reactors. One site has recently been approved, the first in the last 30 years. Both 2007 and 2008 could be critical years for the US nuclear industry. If the licensing and go-ahead are confirmed for one or two nuclear plants, it is likely that several other projects will follow, possibly paving the way for a true nuclear renaissance.

Figure 1.13

New investment decisions needed in United States to lift margins of available capacity to more acceptable levels

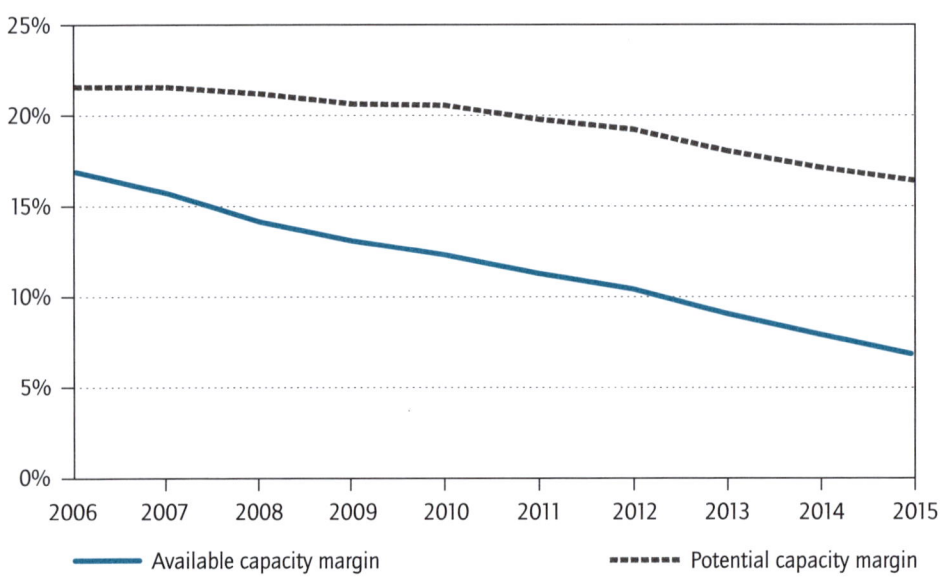

Source: NERC, (2006).

Despite the projected capacity additions, in its recent long-term reliability report, NERC projects that capacity margins in most NERC regions will decline further during the period to 2015 (Figure 1.13). Margins will, of course, vary from region to region but worrisome trends are noted in some cases. For example, without further plant investment decisions, available capacity margins are projected to drop below minimum regional target levels in Texas, the Midwest, New England, and the Rocky Mountains – even in the next two to three years.

Portions of the Northeast, the Southwest and West are expected to reach minimum levels at a later date, but still within the next ten years.

In its Annual Energy Outlook 2007, the US Energy Information Administration projects that in the reference case, 292 GW of new generating capacity will be required by 2030 to meet growth in electricity demand and to replace inefficient, older generating plants that are retired. Coal-fired capacity accounts for about 54% of the total capacity additions expected from 2006 to 2030. Natural-gas-fired plants represent 36% of the projected additions. Renewables account for 6% of the total, and nuclear for the remaining 4%, or approximately 12 GW. Biomass and wind will lead the projected growth in renewable generation. Because fuel costs are a larger share of total expenditures for new natural-gas-fired capacity, higher fuel prices will likely lead to more coal-fired additions. Overall, the largest amounts of new capacity are expected in the Southeast and in the West.

A pilot project is currently underway to develop a capacity adequacy standard for the US Pacific Northwest (see Box 1.2). The methodology proposed highlights the need to establish separate adequacy standards in regions that experience seasonal differences in loads and variable hydro resources. It also suggests taking regional trade flows into account as part of the resource-base.

Box 1.2 . A pilot capacity adequacy standard for the US Pacific Northwest

The Northwest Regional Resource Adequacy Forum has proposed a pilot project to develop a capacity adequacy standard for the US Pacific Northwest region (Northwest Power and Conservation Council, 2006). The Forum plans to test and refine the standard following public consultation, and to recommend a final version by late 2007.

The proposed non-binding guidelines are intended to facilitate the assessment of individual utility resource plans by providing a consistent context to utilities and regulatory entities. In addition, the guidelines are expected to flow into a region-wide assessment of resource adequacy, conducted by the Western Electricity Coordinating Council.

Seasonal differences in both loads and resources make it imperative that the Northwest establish separate winter and summer targets. Resources generally available to the Northwest in winter are desired for meeting summer peaks elsewhere in the west, particularly California. This peak scenario played out during a heat wave in July 2006, when virtually all of Northwest's 3 500 MW uncommitted IPP capacity was transmitted to California, either as direct sales or via resale by Northwest utilities. The pilot capacity standard would lower uncommitted IPP contributions to Northwest summer capacity.

The proposal establishes a 25% planning reserve margin in winter, and 19% in summer. On the resource side, the standard focuses on «sustained peaking capability» from all non-hydro resources and hydroelectricity that can be available under low water conditions. The resources pool also incorporates the net balance of imports and exports by regional firms and seasonal spot-market resources that can be available beyond the Northwest. It also factors in the uncommitted (i.e. no firm contracts) capacity from independent power plants (IPPs) in the Northwest. Wind energy is valued at 15% of its installed capacity.

The proposed reserve margins derive from three sources. First is the anticipated load increase from a 1-in-20-year temperature deviation; the Forum assigned this component a 15% margin for the winter capacity target, and 6% for summer. The second is the operating reserve requirements, which were set at an additional 6% for both summer and winter. Finally, the targets were influenced by additional planning adjustment reserves that are established at 4% for winter and 7% for summer, based on a loss-of-load-probability analysis.

OECD Pacific Tests Roles of Governments and Markets....

OECD Pacific has registered significant capacity additions over the last decade. Electricity demand has been very strong in Korea, while energy efficiency and conservation have constrained demand growth in Japan. Limited demand growth occurred in New Zealand. There are some concerns about resource adequacy in some regions in Australia by 2010. Japan and Korea will likely continue to experience relatively high reserve margins.

■ Australia

Electricity consumption in Australia increased by 3.5% in 2005, with faster growth rates recorded in Victoria and Western Australia. Industry is by far the largest consuming sector. System peak-loads under extreme weather conditions are projected to increase faster than average electricity consumption over the period to 2015. The peak-loads are expected to reach 18 747 MW in New

South Wales and the Australian Capital Territory, 13 300 MW in Queensland, 3 969 MW in South Australia and 12 123 MW in Victoria (NEMMCO, 2006). Generation is predominantly coal-fired (more than 80%), with some gas and hydro. Substantial increases in generation capacity have occurred in recent years, driven by demand growth. In 2005, Australia generated 248.4 TWh of electricity with a total generating capacity of 53 GW.

The establishment, in December 1998, of the National Electricity Market (NEM) marked a key milestone in Australia's energy sector reform. System and market operations are managed by the National Electricity Market Management Company Limited (NEMMCO). In 2005, a single national body, the Australian Energy Regulator, was established and given a mandate to take full responsibility for transmission assets and market monitoring. In 2006, Australia commissioned the Basslink sub-sea connector, physically connecting the island of Tasmania to the NEM.

Figure 1.14

Strong growth in Australia's electricity output still dominated by coal, but importance of gas increasing

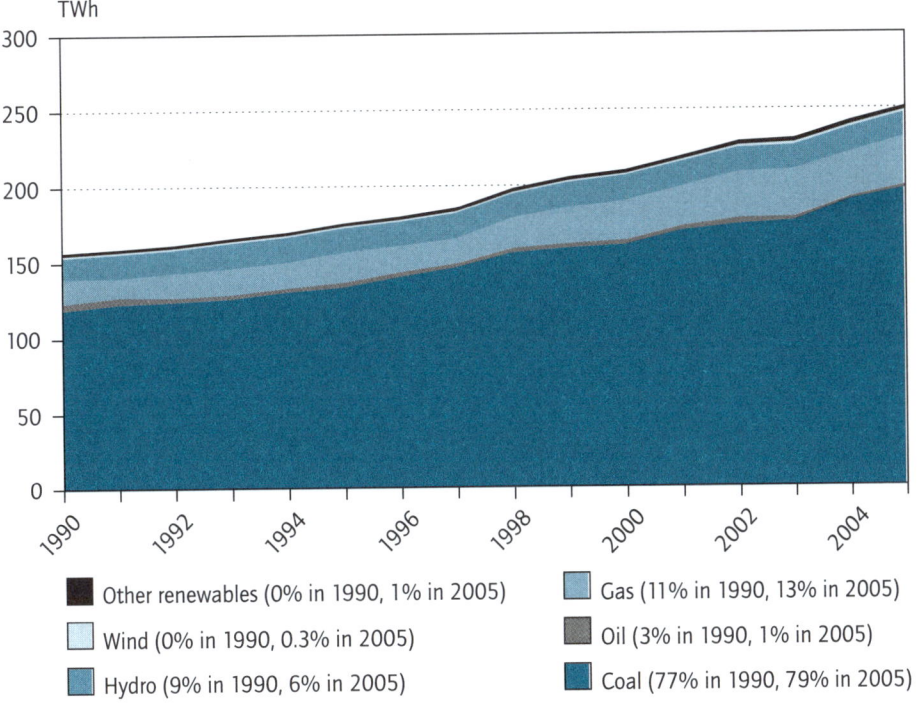

Source: IEA Statistics.

The increase in private generation ownership along with the creation of NEM have encouraged investments in new generation capacity to meet growing demand and compensate for ageing facilities. According to the Platts database (Platts, 2005), 23% of coal-fired power units in Australia are more than 30 years old, compared to 36% for oil and 20% for gas. Between 2000 and 2005, 5.2 GW of new generating capacity was added, mostly gas-fired. The new gas-fired generation mainly meets mid-merit and peak-load. In terms of generation output, Australia's reliance on coal-fired generation has increased (Figure 1.14). Nevertheless, rapid growth in demand for electricity led to declines in reserve margins.

Australia's electricity prices are among the lowest in the OECD countries. Residential prices are 36% of those in Japan and just more than half of European rates. Similarly, industrial prices are 30% of Japanese levels and 60% of European prices. Power tariffs in South Australia and Western Australia

Table 1.1

New power generation developments in Australia's National Electricity Market (NEM)

Project	Company	Status	Comments
Laverton North	Snowy Hydro	Under construction	320 MW, gas-fired
Tomago	Macquarie	State approved 2002	300 MW, gas-fired
Townsville	Australian Gas Light	Expected 2009	370 MW, gas-fired
Woodlawn Wind Farm	–	Approved	50 MW, wind farm
Tallawarra	TRUenergy	Planned	400 MW, gas combined cycle
Braemar	ERM/Babcock & Brown	Planned	450 MW, gas-fired
Wagga Wagga	Babcock & Brown	Originally scheduled for 2006 – delayed	300 MW expansion of gas-fired plant facing government commission of inquiry

Source: NEMMCO.

are higher than the national average, due partly to a higher dependence on relatively more expensive gas-fired power.

Each year, NEMMCO publishes a statement of opportunities (SOO), which provides information about the adequacy of electricity generation and transmission infrastructure as well as the ability of the NEM system to meet projected demand for the next 10 years. In the 2006 SOO report, NEMMCO stated that about 2.2 GW of capacity is under construction and advanced planning. Table 1.1 includes the most significant projects and illustrates the level of detail in the SOO reporting. In order to meet its supply targets, Australia will need to attract USD 5.5 billion of new investment in its transmission and distribution network during the current decade.

A key conclusion of the 2006 SOO report is that without additional capacity, reserve margins could be expected to fall below reliability standards. The report projects that this would occur in the following timeframes: summer of 2007/08 for South Australia, summer of 2008/09 in Victoria, summer of 2009/10 in Queensland, and summer of 2010/11 in New South Wales. Tasmania is better positioned with reserve margins remaining reliable beyond 2015/16. These projections demonstrate the need for additional commitments to new investments to ensure generation adequacy and system reliability.

New Zealand

Electricity markets in New Zealand were liberalised relatively early among IEA member countries, with unbundling completed in 1998. Regulatory structures and market design have changed in recent years as New Zealand gained more experience with liberalisation. The Electricity Commission was established in 2003, with a mandate to monitor and administer the electricity sector.

Electricity generation in New Zealand is predominantly hydro-based (56% in 2005), with gas and coal accounting for approximately 34% of total generation, and geothermal and wind for the remainder. Natural gas-fuelled generation grew from less than 1% in 1980 to 17% by 2004. In 2005, total generation rose marginally from the previous year to 42 TWh. The generation sector is relatively concentrated with five generators accounting for about 90% of total supply. According to the Platts database (Platts, 2005), 28% of oil-fired power units are 30 years old, compared to 21% for gas.

Peak demand has grown from 5.83 GW in 2000 to 6.07 GW in 2004. Over the 2000-2004 period, total capacity increased from 8.15 GW to 8.58 GW, largely as a result of government intervention. Reserve margins improved during the period and stood at about 30% in 2004.

A dozen power plant projects, totalling nearly 1 000 MW, are under construction or planned for 2007 and 2008 start up. These include a 365 MW CCGT, two small hydro projects, three geothermal projects and six wind power projects. Despite the proposed new generation, there is some concern about generation adequacy in the Auckland area, primarily due to transmission constraints. Given its reliance on hydro, the system is constrained by energy rather than by capacity. With demand expected to grow at about 2% per year, the government estimates that new generation to supply 800 GWh is required each year (IEA, 2006a).

Japan

Electricity rates in Japan are among the highest in the developed world, partly because of the country's need to import virtually all of its fuels. High land costs also contribute to push up rates. At the same time, high energy prices have provided strong incentives for energy savings. Markets are served by regional power companies; rates are set separately by each regional company, and are linked to underlying fuel prices and other costs. These rates are generally reviewed on a quarterly basis and the regulated rates for small consumers are subject to approval from METI (Ministry of Economy, Trade and Industry). Japan's electricity prices have trended downwards during the last five years. This trend is attributed to increased competition among power companies, efforts on the part of power companies to implement a "best mix policy" on fuel types and a decrease in investment due to low demand growth.

Large consumers are now able to choose their electricity supplier and to buy electricity directly from a wholesaler or an IPP at individually negotiated rates. Small-scale and residential customers remain obliged to buy power through their single, regional power utility.

The Japanese market is primarily served by privately owned, vertically integrated regional power companies, along with a number of IPPs. Thermal power plants account for approximately 60% of total electricity supply, followed by nuclear generation with 23% and hydropower with 10%. Total generation capacity reached 275 GW in 2006. Japan is the world's third nuclear power producer (after the United States and France) with 48 GW of installed capacity. Overall, according to the Platts database (Platts, 2005), 62% of oil-fired power units in Japan are more than 30 years old, compared to 25% for gas and 14% for coal. About 12% of nuclear power reactors are more than 30 years old.

The nine regional power companies produce three-quarters of the country's electricity. They also control regional transmission and distribution infrastructure. As of April 2005, twenty-one companies formed a new spot wholesale electricity market. The proportion of total electricity traded is, so far, rather low. Competition has been opened to industrial and large customers who make up almost 60% of the retail market.

The temporary reduction of nuclear generation from 2002 due to lower capacity factors was compensated for by oil-fired and coal-fired generation (Figure 1.15). Japan's ageing oil-fired power plants provide reserve capacity to meet peak demand and compensate during maintenance. Coal will likely remain an important fuel for Japan; it could enable the country to reduce its reliance on hydrocarbon imports from the Middle East and minimise its growing liquefied natural gas (LNG) needs. Environmental considerations have been, and will remain, a key challenge for coal-fired generators for the foreseeable future. To address this challenge, the nine regional companies, Japan Power and Central Research Institute of Electric Power Industry (CRIEPI)

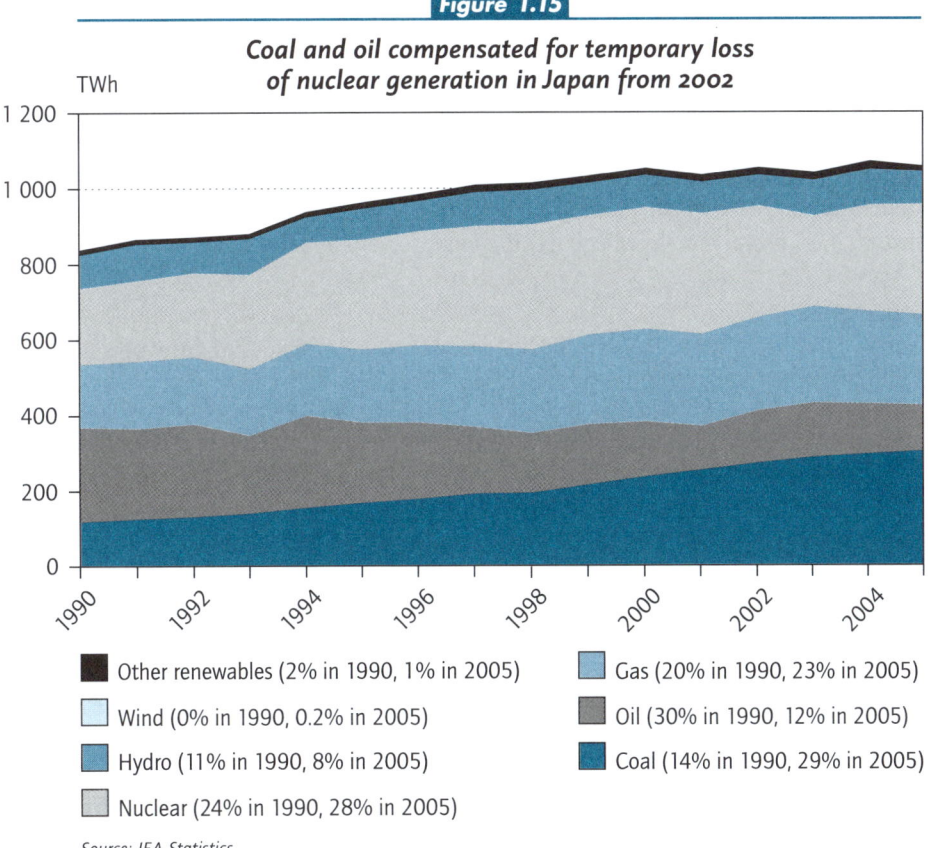

Figure 1.15

Coal and oil compensated for temporary loss of nuclear generation in Japan from 2002

■ Other renewables (2% in 1990, 1% in 2005)
□ Wind (0% in 1990, 0.2% in 2005)
■ Hydro (11% in 1990, 8% in 2005)
□ Nuclear (24% in 1990, 28% in 2005)

■ Gas (20% in 1990, 23% in 2005)
■ Oil (30% in 1990, 12% in 2005)
■ Coal (14% in 1990, 29% in 2005)

Source: IEA Statistics.

established a consortium to build an experimental 250 MW coal gasification power plant. The plant, scheduled for completion in 2007, is expected to achieve a thermal efficiency rate as high as 48%.

Japan currently has a substantial surplus of power capacity, with a peak-load of 174 300 MW in 2004. Most parts of the country have adequate generating margins and reserve capacity to cope with peak-load demand. The network as a whole maintains adequate generation capacity to cope with peak demand, and supply breakdowns are not common. In August 2006, the Nuclear Energy Subcommittee of the Advisory Committee to METI submitted its *Nuclear Energy National Plan.* This will form part of the revised *Basic Energy Plan,* which will be submitted for government approval at the end of 2006/07.

The government objective is to continue to meet at least 30% to 40% of electricity supply even after 2030 by nuclear power generation. The government is now reviewing plans to build around 12 additional reactors over the next decade. It has also set a target of increasing the share of new sources of energy to approximately 3% of total primary energy supply by the 2010 fiscal year. Current Japanese law stipulates that by the 2011 fiscal year, electric power companies must acquire 1.35% of their power supplies from renewable sources. The Ministry of Economy, Trade and Industry has drawn up a new target that would oblige electric power companies to use renewable energy sources to generate electricity worth 1.63%, or 16 TWh, of their projected electricity sales in the 2014 fiscal year. In March 2005, wind power capacity was 926.5 MW.

Korea

Reflecting rapid economic growth, peak electricity demand in Korea has increased at an average rate of nearly 10% per year since 1977, reaching 54 GW in 2005. Although growth in peak demand has tempered recently, it continued to increase at a robust average annual rate of nearly 6% during the past five years. Peak demand is expected to grow by 18% between 2005 and 2013, corresponding to an average annual rate of 2.1%. Ageing facilities do not appear to be an issue in Korea for the period to 2015; only 18% of oil-fired power units are more than 30 years old whereas most coal, gas and nuclear power units are less than 30 years old (Platts, 2005).

By law, the government is required to provide a long-term electricity supply and demand outlook. This is designed to assist the country's six state-owned generation companies in decision making on capacity investment and to ensure reliable electricity supply. Biannually, the government develops a Basic Plan of Long-term Electricity Supply and Demand (BPE). Generation investment in Korea is generally undertaken by the state-owned companies, subject to approval by the Ministry of Commerce, Industry and Energy (MOCIE).

Korea is the world's sixth-largest producer of nuclear power. The 19 nuclear units currently on line account for about 40% of total electricity supply. The government's target is to increase the role of nuclear such that it may account for 60% of the electricity supply mix by 2035.

The six generating companies are now building substantial new capacity; more than 19 GW of new capacity is expected to be on line by 2017 (Figure 1.16). When additions by IPPs are included, the expected new capacity grows to nearly 24 GW, of which 9.6 GW is nuclear and 6.1 GW is coal-fired capacity. In January 2007, Korea announced a major investment in nuclear power, which will boost the contribution of nuclear power to the country's electricity supply mix. The government is planning to spend USD 2.59 billion during the next five years to develop indigenous light-water-reactor technology with the objective to support an export industry in addition to meeting Korea's power needs.

Figure 1.16

Planned capacity additions in Korea add 35% to existing capacity

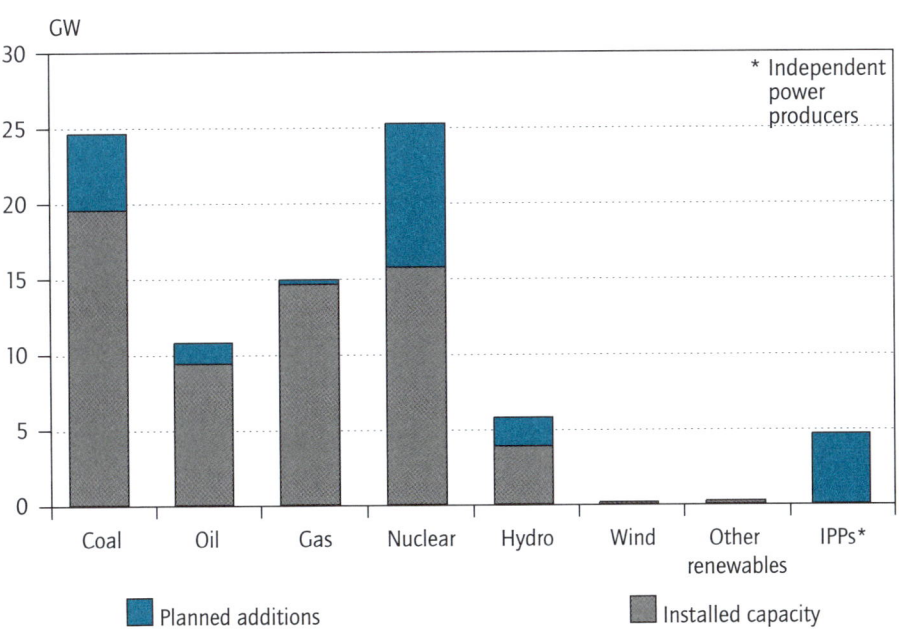

Sources: IEA Statistics and MOCIE.

In recent years, Korea's reserve margin has trended towards 20% (Figure 1.17). Although lower than in the past, it indicates that the country continues to maintain sufficient capacity to meet its growing demand. Generation capacity is expected to grow faster than demand in the coming decade. As a result, the reserve margin is projected to trend upward, reaching 35% in 2013.

Figure 1.17

Volatile margins in Korea (1994-2004) reflected fluctuations in peak demand

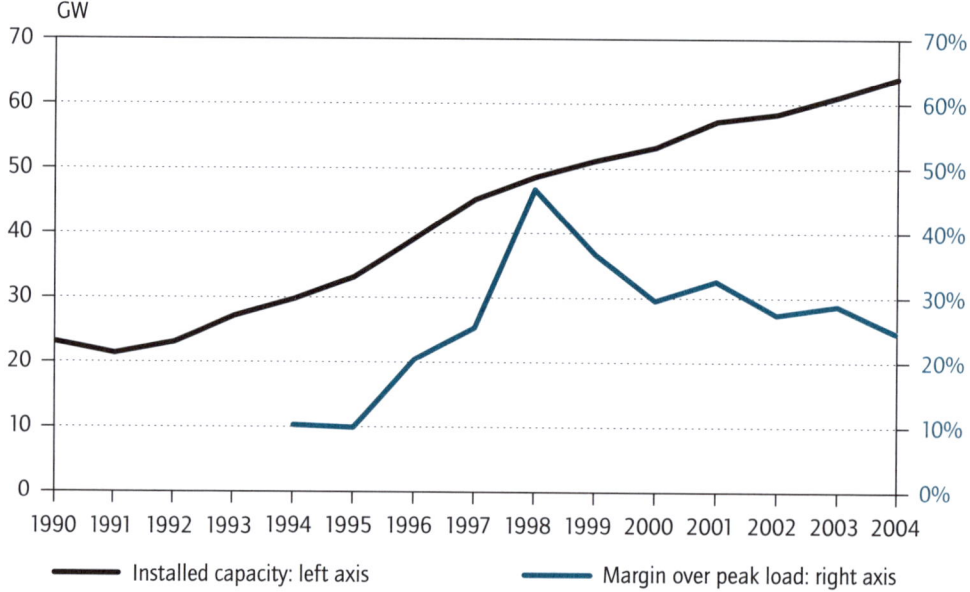

Installed capacity: left axis

Margin over peak load: right axis

DIVERSIFICATION FOR EFFICIENCY, RELIABILITY AND THE ENVIRONMENT

Choice of technology in investment decisions sets the scene for efficiency, reliability and environmental impact. There are no technologies that are unambiguously cheaper, cleaner and more reliable than any other technology in all circumstances. There is a group of more conventional technologies that have strong and weak points in fulfilling these objectives. Considering that investment decisions are made under policy and market uncertainties; diversification is a sensible aspect of the investment response.

Upholding the entire supply chain necessary for reliable and efficient electricity provision is the ultimate objective; diversification in terms of technology and fuel is a sensible strategy to manage all risks. One of the critical risk factors is uncertainty about the direct investment costs, which will determine the profitability of a project and create a strong case for diversification of technologies. Diversification is also called for to manage uncertainty about price and availability of fuel, and policy uncertainty on environmental constraints. There are other more indirect risk components related to the secure operation of the entire electricity system. Some technologies are particularly well suited to contribute to the secure operation of the system. Other technologies tend to aggravate operational challenges, hence, also from an operational point of view there are strong reasons for diversification.

But power generation is only one part of the chain that ensures reliable and competitive supply of electricity to final consumers. Transmission and distribution networks connect power plants with consumers. Transmission systems essentially define the relevant load area for a new generation project. Hence, the profitability of a new project highly depends on the capability and related costs of the transmission system. In that sense, a new transmission interconnector may be another alternative investment opportunity and play a key role in diversification considerations. Transmission effectively competes with generation and in many cases is a direct substitute. If conditions are right, a new transmission line may be a more profitable investment than adding new generation. Appropriate pricing of transmission bottlenecks is crucial for efficient choice of technology, size, timing and location of a new project.

Diversification is thus a sensible strategy for risk management. However, commercial market players will only make such a strategy their own to the extent they can "see" all the real risks. Price is the principal medium for communicating these risks and making them transparent. Market players are responding to market signals – fuel costs, plant costs, and electricity price. Thus, public policy should pursue the objective of making such prices and costs as reflective of reality as possible.

This chapter explores the most important cost drivers in generation investment. The direct cost drivers are the focus of the first section. The next section takes a system-wide view, exploring the need and drivers for diversification in order to uphold the entire chain for reliable electricity supply. The final section focuses on wind power, which involves specific operational challenges due to its intermittency.

Costs and Merits are Balanced Best in a Level Playing Field...

Existing generation technologies represent a wide array of options; all have advantages and drawbacks that weigh differently from project to project. Pulverised steam coal plants, combined cycle gas turbines (CCGTs) and nuclear power are the more generic conventional technologies. On-shore wind power is edging into the ranks of conventional technology with higher and strongly increasing shares, but with performance highly dependant on local wind resources. These four generation types are the focus of the cost analysis in this section.

Hydro power, combined heat and power (CHP), and plants fuelled by lignite, geothermal or biomass are also often highly competitive options. However, they rely heavily on local conditions. Hydro relies on hydro resources, preferably with reservoir storage, although low-head run-of-the-river plants can be viable. To be fully efficient, CHP must have a demand for heat, either industrial or district heating. Lignite plants are highly competitive when located adjacent to a lignite resource, even though the environmental impacts are severe. Biomass (perhaps as a type of CHP) is a natural choice in locations with abundant and cheap supply of biomass, such as proximity to large pulp and paper industries. Old oil-fired plants still constitute an important resource for back-up and new oil-fired plants are the preferred option in some circumstances, such as isolated island systems. A range of distributed generation options, such as solar/photovoltaic and micro generation, are possible competitive choices for the future. They may already be demonstrating benefit in limited and very specific local circumstances. Off-shore wind power is also deployed more frequently, with considerable cost improvements expected.

Levelised Costs

With each generation option, basic costs of generating electricity are the main driver for choice of technology. The key parameters of balanced investment decisions include the main cost components – investment, operation and fuel costs. Main cost components and parameters are listed in Table 2.1 with resulting levelised cost calculations. The levelised cost methodology is the traditional approach for calculating and comparing costs of various generation technologies. It is a calculation of the constant real wholesale price that meets all operating, fuel and financial costs, including debt payment, and income taxes. This methodology is useful for analysing costs at a system-wide level, and for analysing the impact of variations in individual cost factors. For real investment decisions it is, however, only one element in the analysis necessary to understand costs and risks in a specific project. Analysis methods relevant to specific projects are further explored in Boxes 2.3 and 3.2, with a stronger focus on appropriately accounting for real market risks.

The data in Table 2.1 are from *World Energy Outlook 2006 (WEO 2006)* and are based on IEA databases and NEA/IEA (2005)[5]. Many cost components will vary, even considerably, from location to location and project to project. The assumptions underpinning these levelised cost calculations were developed following extensive interaction with market players and experts from all IEA regions.

Levelised costs presented in Table 2.1 are calculated at two different discount rates, reflecting two different levels of cost of capital. They can also be seen as two different assessments of the accepted re-payment time – the time it takes until invested capital is recovered. With a high discount rate, the invested capital is recovered in a shorter time than with a low rate. Capital expenditure for power generation competes with alternatives in global capital markets. Cost of capital can deviate to a certain extent over time. But more importantly, cost of capital for power generation will depend on the relative risk level of a specific investment compared to alternatives. Relative risk level of a project is often reflected in two basic parameters; *i*) The level to which it is possible to finance the project with debt relative to residual equity, and *ii*) The actual rate of return required both on debt and equity. For relatively low-risk projects it may be possible to finance large capital requirements with debt and at low rates. The riskier the project, the higher shares will have to be financed

---

5. *The methodology for computing levelised costs presented in* WEO 2006 *defer from levelised costs presented in IEA/ NEA (2005) on a number of areas. Some additional cost parameters, such as cost-escalation rates, are introduced and the most important difference is the inclusion of corporate tax in* WEO 2006.

Table 2.1

Levelised costs for generation units starting commercial operation in 2015

Parameter	Nuclear[5]	Combined cycle gas turbine	Pulverised coal	Wind - onshore	Open cycle gas turbine
Investment cost[1], USD/kW	2 500	650	1 400	900	400
Construction time, months	60	36	48	18	24
Lifetime, years	40[4]	25	40[4]	20	20
Capacity factor, %	85	85	85	28	1
Thermal efficiency[2], %	33	58	44	-	37
Cost of fuel, USD/MBtu[3]	0.5	6.0	2.2	-	6.0
Operation and Maintenance Costs, USD/kW/year	65	25	50	20	20

Levelised costs, USD/MWh

		Nuclear[5]	Combined cycle gas turbine	Pulverised coal	Wind - onshore	Open cycle gas turbine
discount rate 6.7%[6]	Investment	41	10	22	50	609
	Fuel	7	45	21	0	70
	O&M	9	4	7	9	238
	Total	**57**	**59**	**50**	**59**	**917**
discount rate 9.6%[6]	Investment	65	15	34	65	800
	Fuel	7	45	21	0	70
	O&M	9	3	7	9	237
	Total	**81**	**63**	**62**	**74**	**1 107**

1 Total capital expenditure, excluding financing costs
2 Lower heating value (LHV).
3 Million British Thermal Units (MBtu) is a common unit for natural gas. USD 2.2 /MBtu for coal corresponds to USD 55 /tonne. Nuclear fuel costs include uranium (USD 30 /lbU), enrichment, conversion and fabrication. Fuel price is assumed to escalate at 0.5% per annum.
4 25 years in the high discount rate case.
5 Nuclear costs include USD 350 million for decommissioning and USD 1/MWh for waste disposal.
6 Discount rate: real (2% inflation), after tax (30%), weighted average cost of capital.

Source: IEA, 2006b.

with equity and the higher the required rates of return on equity. Different technologies and projects may be perceived to be of different risk levels.

Perceived risk level also affects the possible corporate form for the ownership of the plant. With relatively low risk and high predictability on basic construction parameters (investment cost, licensing time and construction time) suppliers of debt may accept a corporate form in which the project is assessed in isolation – *i.e.* project finance. Possible losses stay with the plant and are not transferred to its owners – *i.e.* non-recourse. If lenders are not willing to accept such an arrangement, or at least not to a degree and at rates that render the project profitable, project owners will have to assume responsibility for all commitments in the project – financing based on the balance sheet of owners.

Technologies with a good track record and with an expected constant cash flow during their lifetime are regarded as less risky. In contrast, a history of budget overruns, construction delays and fuel cost volatility will add to the perceived risk level. Volatility of demand and other market risks also add to required rates of return.

It is possible to lower the cost of capital in a project by shifting some of the risks to other stakeholders, such as electricity consumers or tax payers, through specific regulatory intervention or subsidies. Such tactics reduce the direct costs of the project but also blur the true underlying risks and can thus distort investment decisions. Policies can be designed to favour any specific development of the generation portfolio, through regulation, interventions and subsidies. But transparency and efficiency have proven often to be lost with such endeavours. Contracts that link generation and electricity consumption will, however, be a natural part of a risk-hedging strategy that can considerably lower costs.

The two discount rates used in Table 2.1 are real after-tax weighted average cost of capital, and are based on a number of financial parameters. Annual inflation is assumed to be 2%. Nominal cost of debt is assumed at 8% in the low discount rate and 10% in the high case. Nominal required rates of return on equity are 12% in the low case and 15% in the high case. Debt shares are 50% in the low case and 40% in the high. Marginal corporate tax rate is assumed at 30%, which is about the OECD average, with 15 years depreciation. The two discount rate cases are realistic examples of how risks can be managed and finance can be organised. The actual risk management and financing opportunities will deviate from technology to technology, from location to location and from project to project.

Total levelised costs of coal, nuclear, CCGT and wind varies greatly with the cost of capital. At the low discount rate, pulverised coal is the cheapest option at USD 50/MWh, with nuclear following behind at USD 57/MWh, and wind and CCGT at USD 59/MWh. At the high discount rate coal is still the cheapest option but is now at USD 62/MW and CCGT at USD 63/MWh. Nuclear power jumps to USD 81/MWh and on-shore wind power to USD 74/MWh.

Levelised costs for open cycle gas turbines (OCGTs) are also presented in Table 2.1 for illustrative purposes. Even though they are far from competitive for base-load generation, they are appropriate to meet peak-load, considering the relatively low investment cost. To demonstrate the specific economics of OCGTs for peak-load, levelised costs are presented at a 1% capacity factor. Levelised costs of OCGTs illustrate the potential impact of price caps. OCGTs intended only to operate for very few hours of the year (1% capacity factor corresponds to 90 hours) have levelised costs at some USD 1 000/MWh. But it is still better than letting more expensive resources stand idle during the other 99% of the time. Increasing the capacity factor to 3% decreases levelised costs to USD 352/MWh in the low discount rate case; at 10% capacity factor, levelised costs are USD 155/MWh. Decreasing the capacity factor to 0.1% (about 9 hours per year) increases levelised costs to about USD 10 000/MWh. Investment in OCGT is particularly risky and probably requires higher rates of return. Increasing required rates of return – to 20% on equity and to 15% on debt – for such investments increases levelised costs to almost USD 1 500/MWh at a 1% capacity factor.

Construction Costs

Investment costs are a critical decision parameter in any investment decision, but its importance varies greatly from technology to technology. The share of investment costs in total levelised costs is 72% for nuclear power in the low rate case and 81% in the high case. Shares are even higher for wind power. For CCGT, the shares are much lower at 16% in the low discount rate case and 24% in the high case. Coal is in between, with shares of some 50%.

Investment costs vary over time and from country to country, and are vulnerable to a number of input factors. Steel is an important ingredient in all generation plants and steel prices fluctuate; at present they are relatively high, driven by high demand. Labour and other specific construction related costs also vary greatly depending on the demand for construction in a specific region. Finally, considering the level of sophistication of all generation technologies, plant vendors are also subject to constraints. High demand puts vendors under capacity pressure. Several IEA countries are faced with an upcoming investment cycle, which is likely to put pressure on the capacity of generation plant vendors, particularly when considering that a considerable demand for new generation capacity from non-IEA member countries must also be met. This adds to the importance of developing competitive electricity markets with effective trade. Effective and competitive markets across large areas will tend to smooth investment cycles in the future.

Plant design standardisation makes it easier to build plants for a larger market and can thus lower costs and help to reduce some supply constraints for vendors. Wind and gas turbines (open and combined cycle) are standardised to a great extent, with many similar plants. Coal plants are adapted to specific local conditions, making standardisation more difficult. Still, investment costs are relatively steady and predictable, building on the long and broad experience of vendors.

Nuclear power is different. The market for new nuclear power plants was dormant in IEA countries for a long time, with Asia being the only exception. Negative experiences with accidents (Three Mile Island and Chernobyl), substantial budget overruns and delays in several countries, and low fossil fuel prices, put new builds on halt for a decade and more (IEA, 2006b). Development and construction experiences in France and Korea are more positive: Standardisation was an important feature and a major contributor to shorter construction times. A high degree of standardisation is a key component of the strategies of nuclear vendors today. Standardisation is pursued with the ambition of bringing down costs, particularly after the first-of-a-kind units. Successful standardisation of nuclear units will be critical for the prospects of new nuclear power. It will facilitate licensing, supply of plant parts and final construction – all of which are crucial to reducing costs.

The sensitivity of investment costs and discount rates, particularly for nuclear power, are illustrated in Figure 2.1. Levelised costs are computed with four different assumptions on investment costs and construction time.

Actual and estimated investment costs for nuclear plants range across a wide span. NEA/IEA (2005) report of projected investment costs of nuclear projects in the Czech Republic and Korea at some USD 1 500/kW (2006 prices). At the other end of the scale is a project in Japan at USD 2 500/kW (2006 prices). Latest data on the Finnish project in Olkilouto suggests some USD 2 600/kW. A recently announced project in Bulgaria suggests a cost of USD 2 550/kW (Platts, 2006). The span relates both to the generic costs of the plant but also to local construction cost features, such as local labour costs and design requirements. Figure 2.1 illustrates clearly that the actual investment cost plays a critical role in the overall competitiveness of the plant, a role which increases rapidly with the level of the discount rate. The difference in levelised costs between a plant at USD 2 000/kW and a plant at USD 2 500/kW is USD 8-10/MWh – which may very well be the difference between being profitable and operating at a loss. Using investment costs at USD 2000/kW in the low discount rate case reduces levelised costs to USD 49/MWh, which is even lower than pulverised coal. Uncertainty about actual investment costs is one of the greatest risks that a nuclear project is facing, and experiences with the first new builds in USA and Europe will be critical for the future prospects of nuclear power in these regions.

Figure 2.1

Levelised costs of nuclear power are very sensitive to investment costs, discount rate and construction time

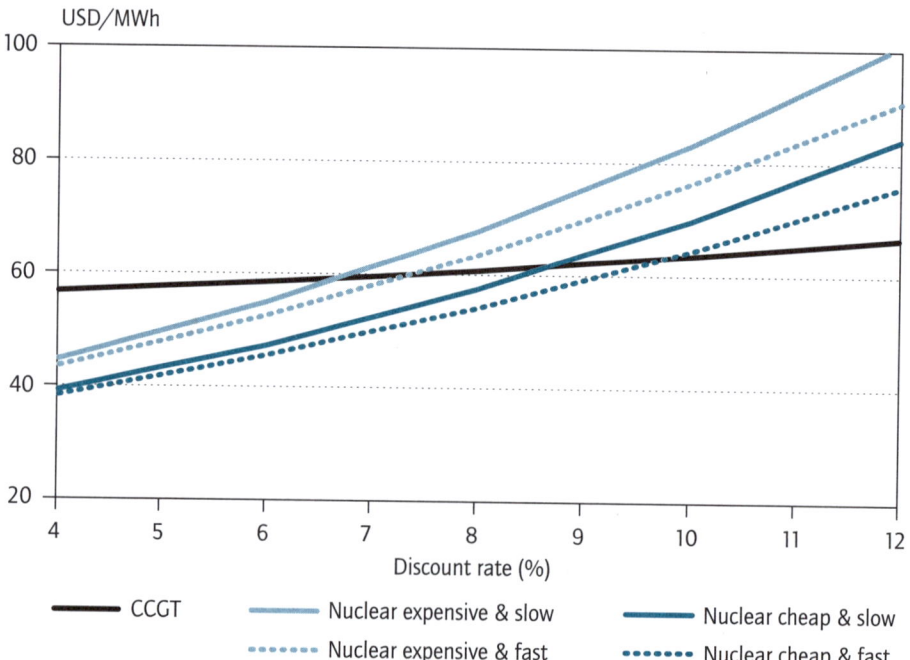

Nuclear cheap = USD 2000/kW, Nuclear expensive = USD 2500/kW, Nuclear fast = 4 years construction, Nuclear slow = 6 years construction

Source: IEA

High up-front investment costs also make construction time of a nuclear plant an important factor. Considerable budget overruns and substantial delays created a problematic track record for nuclear power in many countries. Figure 2.1 illustrates that construction time does matter: a 12 month delay adds 3% to the total levelised costs. Vendors look for options to shorten construction times while benefiting from experience of the first new builds, which could prove an important contribution to lower costs. However, past experience illustrates that the main reason for delays was not due to engineering and actual construction problems. Rather they stem from licensing and approval delays, partly caused by public protests.

Wind and gas turbines are at the opposite end of the scale. Wind turbines are constructed in 6 to 24 months, depending on the size of the total project. OCGTs are assumed to be built in 24 months in the levelised cost calculations, but can often be built in six months. It is possible to add a combined cycle as a module. CCGTs are built as quickly as 18 months in ideal circumstances, but can take up to 36 months and longer.

Unit size is another factor related to investment costs that greatly affects the risk of the project. Today's state-of-the-art nuclear reactor designs range between 1 000 MW (such as the AP1000) and up to 1 600 MW (the EPR under construction in Finland). Units of 1 GW or more may have a great effect on the supply and demand balance, particularly in smaller electricity systems. Demand is increasing slowly or even stagnating in many IEA countries, and a boost of supply with 1 GW or more may drive the electricity price down below the necessary level for profitability, at least for a period of time.

Nuclear power possesses important economies of scale both in the size of each unit and in the number of units. Licensing, construction, operation, safety management and waste management are cheaper and more efficient with a group of nuclear plants compared to one or very few. The United States provides a good example of the importance of scale in nuclear power. Ownership of nuclear power plants in the United States has experienced one of the most significant consolidations in the US electricity industry, which may still not have come to an end. The United States now has 28 different operators; the 6 largest operate 50% of installed capacity but the 8 smallest operate less than 2 GW each. In 1991, 101 different utilities were involved. Again, CCGTs and on-shore wind power are at the other end of the scale. Both these technologies can be built in relatively small sizes without significantly increasing cost per kW of installed capacity. CCGTs can thus be built in stages, commissioning the gas turbine before the entire plant, and in modules, increasing the capacity in steps of 300-800 MW. Coal units are typically built in unit sizes of 300 MW to 1 000 MW. There are thus important economies of scale, but less so than in nuclear power.

In this report, levelised costs are calculated for generation projects to be commissioned by 2015. Extensive R&D efforts by commercial firms and governments are helping to broaden the pool of technologies to choose from for cheap, reliable and clean power generation. Extensive efforts are undertaken to make break-throughs in next generation nuclear power and several renewable technologies. Technologies to reduce the environmental impact of coal power are also receiving greater attention, primarily because of the prospects to realise significant advancements in climate change abatement on a somewhat shorter time scale through evolution and deployment. Box 2.1 further explores the prospects for clean coal technologies.

Box 2.1 . Clean coal technologies under development

Coal resources are abundant and distributed across many countries. Coal-fired generation is often competitive with alternatives and will be an indispensable part of the generation mix in most IEA member countries for the foreseeable future. But burning coal in power stations is one of the most significant contributors to the emissions of greenhouse gases. Higher efficiency power generation from coal may well be a relatively cost-effective option to reduce emissions and may result, in any event, with evolutionary technical developments. According to a recent IEA study, Energy Technology Perspectives (IEA, 2006c), average coal plant efficiencies have increased from 34% in 1970 to 37% in 2003 in the United States. In Western Europe, the increase is from 32% to 39%; in Japan from 25% to 42%. In Table 2.1 it is assumed that a standard pulverised coal plant will achieve 44% efficiency in 2015, corresponding to today's better supercritical pulverised coal plants. Improvements beyond that come at a cost, but efficiencies as high as 50% and even 55% are within reach (IEA, 2006c). There are two main avenues to reach that goal: ultra-supercritical pulverised coal plants and integrated gasification combined cycles (IGCC).

In ultra-supercritical plants, high steam temperatures and pressures allow more of the heat of combustion to be converted into useful electrical energy; less heat is wasted to the cooling water and flue gases. However, more expensive materials are required to handle the high temperatures and pressures. Some industry estimates now indicate that the boiler and steam-turbine in an ultra-supercritical plant cost about 15% more than in a commercial supercritical plant, depending on the targeted efficiency. The impact of this higher capital cost on the overall plant economics is, to an extent, balanced by the increased efficiency, which brings cost savings in both fuel and fuel handling. A CO_2 price would push investors towards higher efficiency plants, assuming that coal remained competitive with lower carbon alternatives. Commercial ultra-supercritical plants are in operation in Japan, Germany and Denmark with efficiencies just below 50%. To achieve efficiencies of 50% and above requires further improvements in materials.

IGCC is another approach to increase efficiency. Higher working temperatures are achieved through gasification of the fuel, which is then directly burned in a gas turbine. Gasification technology can process any carbonaceous fuel, including coal, petroleum coke, residual oil, biomass and municipal solid waste. This flexibility can be an advantage compared to pulverised coal plants. Providing that natural gas is available, it is also possible to use an IGCC plant as a normal CCGT, thus adding even further fuel flexibility. Several demonstration plants are now operating in Europe and the United States; one is under construction in Japan and others planned in China. Current costs are 20% higher than for pulverised coal plants, although large plant vendors plan to launch new plant designs in the near future also with the objective to push down costs (IEA, 2006c).

CO_2 capture and storage deep underground can significantly reduce CO_2 emissions, by 90% or more. This option is still under development, with important R&D and demonstration still required before it is ready for commercialisation. The additional CO_2 capture process, as well as transport and storage, add costs for which investors would require a return, earned from the CO_2 abated. The capture process reduces the overall efficiency of power generation and significant new infrastructure would be required to transport CO_2. It is possible to use relatively small quantities of CO_2 to enhance oil recovery in oil fields, potentially bringing a value to CO_2 storage at particular locations. Some industry-led projects have demonstrated this as an early CO_2 abatement opportunity. If CO_2 capture and storage becomes an established mitigation measure, then capture from an IGCC plant is technically easier than post-combustion capture from a conventional steam plant. The former is demonstrated at scale, the latter is not. Prospects for carbon capture and storage are explored in detail in IEA studies (IEA, 2004b and IEA, 2006c).

In reality, actual construction represents only the final phase in a project that starts several years in advance. A project also needs planning and development and a long range of licenses and approvals is required, all varying with project, location and technology. Again, nuclear power projects require the longest pre-construction process, a process that also necessitates considerable investments before even knowing that the project will be realised. A total pre-construction cost of USD 200 million has been mentioned by commentators in the UK market (Robson, 2006). Public acceptability of a project is reflected in this process, and is a real threat to any generation investment project. Licensing and other regulatory aspects are explored further in chapter 4.

Fuel and Other Operational Costs

Fuel costs completely reverse the overall cost picture. Wind has no fuel costs, and this is one of the main competitive advantages of this technology. For nuclear power, fuel costs only represent a small share (between 8% and 11%, depending on discount rate) of total levelised costs. About half of the nuclear fuel costs accrue to enrichment, conversion and fabrication, costs which are relatively fixed. Even a doubling of raw nuclear fuel costs would only add some USD 2-4/MWh to the total levelised costs. CCGTs are, on the other hand, very sensitive to fuel costs. Fuel costs represent some 75% of total levelised costs of CCGTs. Considering that the price of natural gas tends to be very volatile, this is an important drawback for CCGTs. Coal-fired plants are also more sensitive to fuel costs than nuclear power, but less so than CCGTs. Figure 2.2 illustrates the sensitivity of levelised costs to variations in coal and gas prices based on a discount rate at 6.7%.

Coal and gas prices assumed in the levelised cost calculations in Table 2.1 correspond to the price projections in WEO 2006 towards 2015 (IEA, 2006b). In the low discount rate case, CCGTs have lower levelised costs than nuclear at gas prices below USD 5.8/MBtu. Relative competitiveness of coal and CCGTs depend on the relative prices of coal and gas. Coal remains more competitive than nuclear until coal prices rise above USD 70/tonne. Markets for natural gas, coal and uranium are explored further in the next section, including historical price developments.

The combination of high fuel cost coupled with low investment cost improves the actual market situation for CCGTs. Investment costs in a power generation plant can be considered "sunk costs" from the moment they are incurred. From the moment a plant is commissioned, the marginal cost of producing an additional unit of electricity should determine its operation (dispatch).

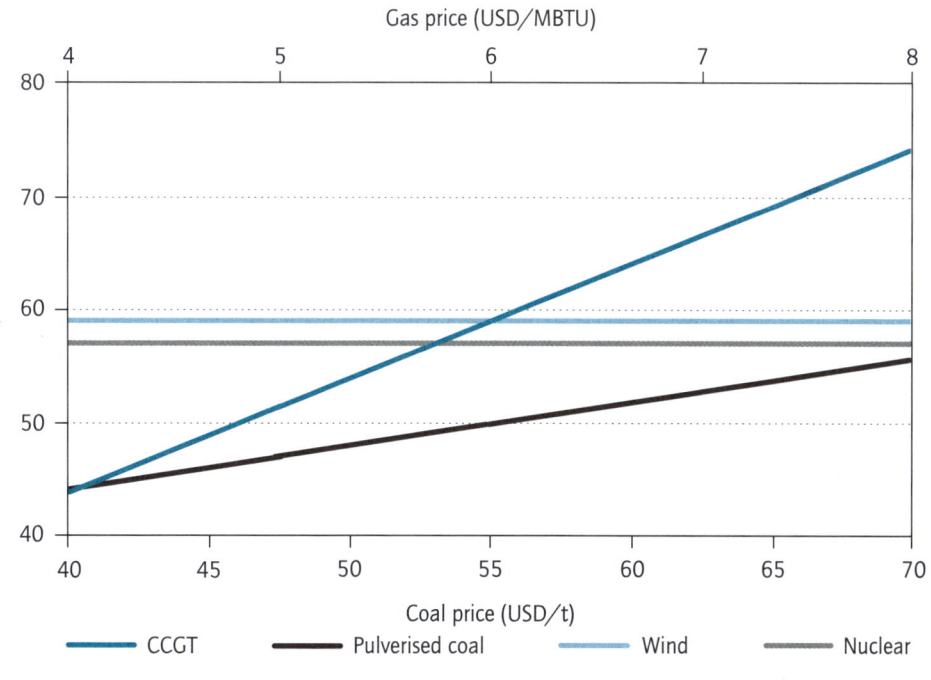

Figure 2.2

Competitiveness of gas-fired generation is highly sensitive to fuel prices

Source: IEA.

Marginal costs more or less correspond to fuel costs; thus, CCGTs often have the highest marginal costs, even at relatively low gas prices. In many cases CCGTs are the marginal plants that determine the price in competitive markets. Hence, increases in gas prices are passed on as increases in wholesale electricity prices. CCGTs still have most of their costs covered even if gas prices increase. High gas prices make other alternative technologies the most competitive, but CCGTs may still be perceived as the least risky. In other words, CCGTs are less financially vulnerable to being left out of the dispatch, due to the relatively low investment costs.

Operation and maintenance (O&M) costs are the third most important cost component. For nuclear, coal and wind power, O&M costs are significant, varying between 10% and 15% of total levelised costs, also depending on the discount rate. O&M costs for CCGTs are considerably lower.

The actual costs of maintaining generation plants are important – but perhaps even more important is the cost in terms of time off the grid. The higher the investment costs, the more important it is that a unit operates as many hours as possible – *i.e.* to spread investment costs over as large a generation output as possible. Levelised cost calculations in Table 2.1 are based on a generic load factor of 85% for nuclear, coal and CCGTs. Figure 2.3 shows the relative sensitivity of levelised costs to the capacity factors in the low discount rate case.

The impact of capacity factors on levelised costs is highly dependent on the share of investment costs of total cost. Nuclear power has lower levelised cost than CCGT at capacity factors above 80% in the low discount rate case. CCGTs are the most competitive at low capacity factors, below 55%.

Figure 2.3

Competitiveness of nuclear power is highly sensitive to capacity factors

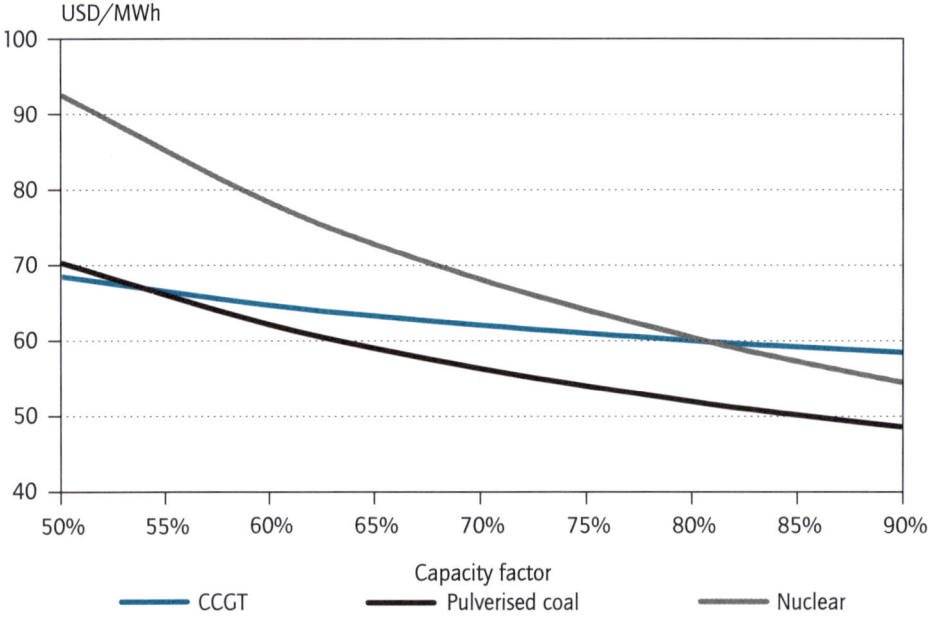

Source: IEA.

Performance of nuclear power plants has improved significantly in most countries. Prior to 1985 in the United States, average capacity factors of nuclear plants were at 50-55%. A decade later they had increased to some 80-85% and today they are at 90% or higher. Finland and Korea stand out with remarkably high average capacity factors at 95% and above. Other important nuclear countries, such as France and Japan, also experienced improved capacity factors reaching 80% and above. There was a dip in

Japanese average nuclear capacity factors from 2002 due to forced outages in connection with incidents relating to safety regulation. Interestingly, in France, average capacity factors were high during the first years of operation, until the second half of the 1980s. From then, nuclear capacity in France increased rapidly, from 35% to 55% of total installed capacity. Average capacity factors dropped with the increase in capacity, only to recover again in later years. This is related to the increased possibility to export electricity rather than the actual availability and performance of French nuclear power plants.

Average capacity factors reflect the amount of time a unit is able to operate at full capacity. In reality, this is a reflection of both operational performance and of market risk or market performance. A unit with high investment costs is suitable for base-load supply, but if the total amount of installed base-load capacity is higher than minimum demand in the entire system and maximum export capacity, some base-load capacity is forced out of the market during some periods. This is a considerable risk with a high share of base-load capacity. Market risk and sensitivities to capacity factors, in combination, single out CCGT as the logical choice to supply mid-merit load. This added to the reasons for building CCGTs during the last decade; in addition to low gas prices generation portfolios in several countries could also benefit from incorporating this technology in the generation mix to meet mid-merit load.

Wind power is a different story. First of all, potential average capacity factors of wind power rely on wind resources at the specific locations of wind turbines, even if outage time in connection with maintenance also plays an important role. Increasing average capacity factors of wind power from 28% to 32% bring levelised costs down to just above USD 50/MWh in the low discount rate case. This is well below nuclear and CCGTs. Capacity factors at 28%, 32% and even up to 35% are achievable at good on-shore wind sites. Capacity factors off-shore may be significantly higher, but still not high enough to compensate for the increased investment and O&M costs of off-shore plants. Average capacity factors in IEA countries are significantly lower than 28%, reflecting great variety in the quality of wind sites used (IEA, 2006f).

Environmental Costs

Investment costs, fuel costs, O&M costs, financing costs and plant performance are the most important direct cost drivers that influence investment in generation. The importance of indirect costs is now recognised by all IEA countries, particularly those costs associated with environmental impacts that result from CO_2 and other greenhouse gas emissions. The "burning" of wind and uranium produces no CO_2 emissions. Coal has the highest carbon intensity, about 75% higher than natural gas. This difference is reinforced through the combustion of the fuel; CCGTs have higher efficiencies than coal plants. Figure 2.4 shows the sensitivity of relative competitiveness of nuclear, wind, CCGT and coal to prices of CO_2 based on the low discount rate case[6]. CO_2 prices or costs are explicit in the European Union (EU) with the introduction of the European Emission Trading Scheme (EU ETS) in 2005. Most IEA countries have various subsidy schemes that function as compensation for non-emitting technologies, particularly for wind power and most other renewable technologies but also for nuclear power in some cases.

Figure 2.4

Relative competitiveness of CCGT, pulverised coal, wind and nuclear changes completely at USD 10-20/t CO_2

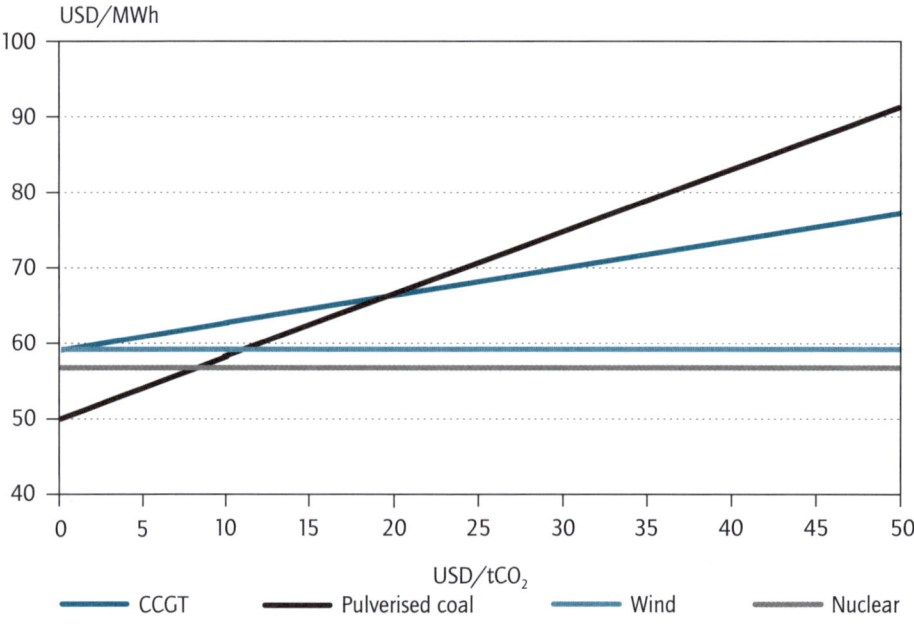

Source: IEA

Without a price on CO_2 pulverised coal has the lowest levelised costs in the low discount rate case and wind power is at par with CCGT. Coal becomes more expensive than nuclear at USD 8/tCO_2 and wind at USD 12/tCO_2. At

---

6. CO_2 intensity of coal is assumed at 48.96/MWh fuel input. For gas it is assumed to be 28.26/MWh fuel input.

USD 20/tCO$_2$, coal and CCGT start competing. CO$_2$ prices in the EU ETS have fluctuated between USD 1 and 40/tCO$_2$, reflecting great uncertainty, but also reflecting competition between gas and coal, depending on the relative prices of gas and coal.

Hydro power, CHP, biomass and distributed generation also have clear advantages over coal and gas-fired plants in terms of CO$_2$ emissions. These technologies may have competitively low levelised costs when beneficial conditions support their use (access to hydro reservoirs, heat demand or cheap biomass). Hence, there is a pool of potential technologies to choose from and a price on CO$_2$ emissions can fundamentally change investment decisions. However, excluding nuclear power as an option reduces real generation options considerably. If local conditions are not favourable, the only option to reduce CO$_2$ emissions may be to shift from coal to gas.

Decommissioning costs at USD 350 millions per reactor are included in the levelised cost calculations in Table 2.1. These are assumed to accumulate over the first 20 years of the lifetime of the plant. In addition, USD 1/MWh are assumed to be set aside for waste management. Experiences with decommissioning costs and the practices used in several OECD countries are explored in NEA (2003). Reported decommissioning costs of existing plants are in the USD 200 to 500/kW range but climb as high as USD 2 600/kW for some gas-cooled reactors in the United Kingdom (IEA, 2006b).

Decommissioning and waste management do not have significant direct impacts on levelised costs of nuclear power. Their importance is more in terms of the risks that follow from lack of clear policy on these areas. Lack of government decision and regulation on decommissioning and waste management, particularly final disposal of high radioactive waste, is likely to have an important negative impact on public acceptability of new reactors. Required funds to manage decommissioning and waste are not significant if collected through the lifetime of the plant. But inadequate and delayed collection of funds, or inadequate ring-fencing of funds, create a considerable financial burden at the end of its lifetime. The prospects for nuclear power are explored further in Box 2.2 (page 90).

Diversification to Manage Risks

Levelised costs reveal important insights into the main cost factors of alternative generation options. Sensitivity analysis provides a good first-hand impression of the relative sensitivities of the various cost drivers. Levelised costs are, thus, an important tool for policy makers in understanding the main cost drivers of an electricity system and in assessing the importance of

Box 2.2 . Nuclear power at a turning point?

Nuclear power has levelised costs in a range that makes it competitive with gas-fired generation as base-load. It is also competitive with coal if only a relatively low CO_2 emission cost is internalised. Rising gas prices and increasing focus on climate change in recent times have considerably changed the basic economics of nuclear power, thereby contributing to its possible revival. Without such a revival, even at limited scale, the share of nuclear power in the energy mix will decrease towards 2030 in OECD countries.

Several major obstacles must be overcome in order for a revival to materialise. On the technical side, the nuclear industry must re-establish a track-record of project completions on time and within budget. On the financing side, contract and financing models must be developed to cope with the market risks that have become transparent with liberalisation of electricity markets. Investments with long lead-times and high investment costs, such as nuclear power, tend to be seen as too risky in an environment of great uncertainty, including uncertainty during a liberalisation process. Maintaining the role of nuclear power will require urgent government action to reduce regulatory uncertainty and to establish a framework where incentives are clear and strong. This will allow investors to take a long-term view.

Policy on nuclear power is determined largely by public support. It varies from country to country, but the common drivers are considerations about costs, climate change, energy security, safety, physical security, nuclear proliferation, long-term disposal of high radioactive waste and decommissioning. Experience shows that governments must determine clear frameworks on all or most of these issues to ensure the viability of nuclear power. Governments must send strong signals to investors that effectively reduce policy and regulatory uncertainty.

Processes for design licensing and plant approval determine the security and safety regime, and have a considerable direct impact on costs. Putting a price on CO_2 creates incentives for investment in cleaner technologies to abate climate change. Some countries also provide a premium or subsidy to non-emitting technologies. Costs for waste disposal and decommissioning are determined by government decisions. Liberalisation of electricity markets makes risks more transparent and imposes healthy competitive pressure, even in the nuclear power sector. In addition, well-

functioning, deep and liquid markets improve the scope for effective risk management, which may make it easier to attract financing for nuclear power within such a competitive framework. Governments play an important role in the development of such markets. In short, government support is required to keep the nuclear option open and may best be achieved by clearly defining the basic framework conditions. Nuclear power will only become more important if the governments of countries where nuclear power is acceptable play a stronger role in facilitating private investment, especially in liberalised markets.

Even if many factors support the case for a revival of nuclear power, the speed and force of such a revival should not be overstated. Lead times are long for planning, licensing, and for construction. For countries considering adopting the nuclear option for the first time, the time required to build necessary regulatory institutions must be added. The first new plants will be a test of modern designs and of public acceptance. The long-term role of nuclear power will be tested over the next 10 to 20 years.

Policies on nuclear power are going through major changes in several IEA countries. The US Energy Policy Act of 2005 *marked one of the final steps in a policy development to facilitate the building of new nuclear power plants. It includes an innovative support scheme for overcoming the particular uncertainty of the first new-builds, particularly by providing financial insurance for licensing delays. (Nuclear power in the* US Energy Policy Act of 2005 *is explored further in Box 4.2.) The Japanese government issued a Nuclear Power National Plan in 2006, which is part of the New National Energy Strategy. It is to continue to meet at least 30% to 40% of electricity supply even after 2030 by nuclear power generation, within the context of liberalisation of the Japanese electricity market. The UK government issued a comprehensive Energy Review in 2006 that identifies new nuclear power as one of the possible important means to meet future energy challenges in the United Kingdom. Future policy is to facilitate the construction of new nuclear power by the private sector in competition with alternatives. IEA (2006c) gives an overview of the most recent developments in the nuclear industry and in policy.*

policies on issues such as climate change. Levelised costs may also provide some insight for investors in a first screening of generation options. For real investment projects, levelised costs are, however, only a small part of the full investment analysis. The analysis must also include a comprehensive risk analysis, in which multiple risks are taken into account. Box 2.3 explores one methodology for analysis of multiple risks based on simulations of expected net present values (NPV) of a project.

Box 2.3 . Analysing multiple risks in investment projects

Cost calculations in this chapter are based on the levelised lifetime cost approach, in which cost cash flows are discounted back to the present. This calculation determines the costs that will have to be compensated by payment from electricity consumers in a world without uncertainty. The levelised lifetime cost approach is an important part of the analysis of generation costs. However, the methodology poorly reflects the multiple uncertainties and risks for investors involved in a real project.

A more accurate assessment of investment projects can be obtained by using alternative calculation and analysis models that are better able to reflect multiple risks in the marketplace. One method is based on calculation and analysis of the net present values (NPV) of alternative projects. The NPV method calculates the net present value of all cash flows in a project, including revenues, with the same types of assumptions as those made in the levelised lifetime-cost approach. The difference is that NPV focuses directly on the profitability of a project instead of only its cost, thereby introducing electricity price into the equation. The NPV method allows for simulations in which multiple uncertainties and risk factors are taken into account. The assumptions about the possible and expected outcomes of the various cost factors will determine the possible and expected outcomes of NPVs.

The IEA conducted a so-called Monte Carlo simulation to demonstrate how such an approach can provide additional insights to investors and industry planners about the impact of technical, operational, and price risk, as compared to the levelised cost methodology. The exercise included 100 000 simulation runs, based on the technical and cost parameters in Table 2.1 (without corporate tax and using a 5% and 10% real discount rate), and with an assumed electricity price of USD 70/MWh. All technical parameters[7] were modelled by a triangular probability distribution with

7. Investment costs, construction time, O&M costs and capacity factors.

lower and upper bounds at ±20% around the base parameters. Triangular probability distributions for fuel and electricity prices were assumed to vary ±50% around the base assumptions. Figure 2.5 illustrates the modelled NPV cumulative probability distributions. The 50% probability marks the most likely – the expected – NPV. Probabilities below 50% illustrate the distribution of possible NPV outcomes that are worse than the expected. Probabilities above 50% illustrate the distribution of outcomes that are better than expected.

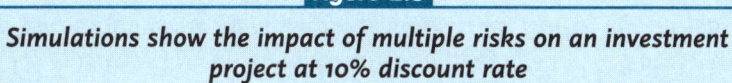

Figure 2.5

Simulations show the impact of multiple risks on an investment project at 10% discount rate

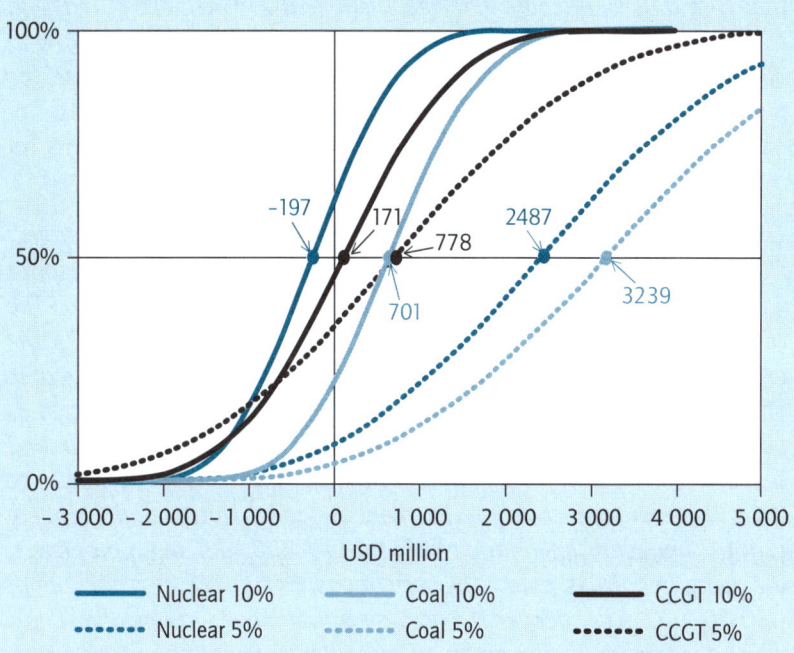

USD million

— Nuclear 10%   — Coal 10%   — CCGT 10%
...... Nuclear 5%   ...... Coal 5%   ...... CCGT 5%

Source: IEA.

At an electricity price of USD 70/MWh and a 10% real discount rate before tax, the expected NPV for coal is USD 701 million, for CCGT it is USD 171 million and a nuclear project has a negative NPV at USD 197 million under these assumptions. At a 5% discount rate, the expected NPVs are all positive: USD 3.2 billion for coal, USD 2.5 billion for nuclear and USD 0.8 billion for CCGTs. The investment rule is that projects with a positive NPV will be profitable and the alternative with the highest NPV should be preferred. When comparing these results with Table 2.1 one should beware that the impact of corporate tax is not taken into account. The introduction of risks does not change the relative ranking of the three technologies in this example compared to Table 2.1 but the differences are more marked. Whereas coal and CCGT have levelised costs at almost similar levels, the introduction of the particular risks used in this example makes coal a far more attractive alternative than CCGT. This approach provides an investor with rich insight on the impact on profitability from multiple variations and risk in key technical, operational and cost parameters. This example shows that with a 10% discount rate there is only a 21% chance of realising a loss with a coal plant. With a CCGT the probability is 44%; with nuclear it climbs to a probability of 58%. At a 5% discount rate the distribution is more flat resulting in small probabilities of a loss for nuclear and coal plants and still about a 30% probability of a loss for a CCGT project.

The slope of the cumulated probabilities of NPVs for the CCGT project illustrates that, in an environment of high fuel and electricity price variations, there are chances of greater losses and greater gains. The cumulated probability curve is flatter for the CCGT project than for the other projects, showing that the spread of possible NPVs is greater. This result does not properly reflect one important effect, which is particularly crucial for CCGTs: Plants can and will close down and sell the fuel back to the market when the electricity price does not cover the marginal costs of the plant. Roques, Nuttal and Newbery (2006) conduct a similar analysis but with slightly different parameters, the most important being the introduction of a CO_2 price of about USD 40/tCO_2. They deepen the analysis by adding operational flexibility, which they define as the ability to close down when the price is below marginal costs. Such operational flexibility considerably increases the expected NPV for the CCGT project and slightly raises the expected NPV for the coal project. With their assumptions, the expected NPV of the CCGT project increases from about USD 400 million to USD 800 million with a 10% discount rate.

The methodology of NPVs and expected NPVs in Monte Carlo simulations was explored further in IEA/NEA (2005).

To summarise, levelised costs of nuclear, coal, gas and wind units are within a range that makes them all highly relevant to consider as a part of a generation portfolio. In fact, a well-balanced generation portfolio – taking all costs, benefits and risks into account – should probably include all these technologies and perhaps others as well, depending on local circumstances. Changes of fuel costs, discount rates, construction costs and CO_2 costs – even within ranges that are realistic and have been observed in the marketplace – fundamentally affect the relative competitiveness of these technologies. CCGTs stand out as a technology offering the most flexibility and lowest risk, even considering the volatility of gas prices. Nuclear power stands out as a technology that offers an impressive economic up-side if the circumstances are right and construction goes well. However, the overall economics are also very sensitive to changes in some critical cost factors. This sensitivity can easily increase the tendency of investments with large up-front investment requirements to be seen as too risky for commercial investors. Government action is urgently needed to reduce uncertainty and to provide a framework with strong and clear incentives to help lower and manage investment risks. Coal plants have the lowest costs under most circumstances except one – a price on CO_2. Advancements on carbon capture and storage technologies are necessary to maintain cost competitiveness of coal in an environment of serious carbon constraints.

Key Message

All available generation technologies have a role to play in cost-effective, reliable and environmentally responsible electricity systems.

Generic generation costs highlight important decision parameters, but also omit many equally important parameters that change with every generation project. Governments need to establish a level playing field to nurture the development of an appropriate generation mix. An increasing role of technologies with lower CO_2 footprints will require government decisions on long-term emission constraints. Nuclear power can play a more important role, but will require government involvement to define a workable regulatory framework with clear incentives and, not insignificantly, to engage in the necessary public debate.

Diversification Needed to Secure Efficiency and Reliability

Reliability of electricity supply depends on an unbroken chain that stretches from the gas field, coal or uranium mine to the electrical outlet in peoples' homes. This supply chain consists of three main elements: energy security, adequacy of assets and system security. Supply of fuels must be secure. There must be adequate assets to convert fuels into electricity and to transport the electricity. Finally, the system must be operated securely to balance supply and demand in real time. This book focuses on power generation, but it is intricately linked to all aspects of the supply chain. No single generation technology possesses all the necessary features to perform all required services in the most efficient and secure way. Fuel markets are volatile and secure system balancing puts extra demands on at least some of the generation assets. Efficiency and reliability calls for a diversified generation portfolio. Adding the environmental dimension further emphasises the need for technological diversification and development.

An efficiently diversified generation portfolio also takes into account the basic plant economics. The total volume of plants with high investment costs should not cover more than the minimum load (and export capacity) of the system. This allows them to operate as base-load plants and recover costs in as many hours as possible. Less expensive plants should cover the additional load in hours when load is higher – mid-load – even if marginal costs are higher. The same is the case for the few peak-load hours of the year, at which time OCGTs are one of the appropriate technologies, even if marginal costs are very high.

There are many reasons to diversify. Commercial market players respond to the needs of the system by adding up the things they see as relevant for profitability – plant costs, fuel costs, electricity prices, etc. From the public policy perspective, the main question is whether governments need to take an active role in order to improve the choice of generation technology by commercial market players. *Are there important factors that commercial market players ignore or do not see? Factors that governments are aware of and can influence through clear policy?* Security of supply is sometimes discussed in the context of public goods. A public good is characterised by having no or inadequate private economic incentives to supply the good. Environment, defence, health care and education are examples of goods that are often fully or partly regarded as public goods. The question is to what extent parallels can be drawn to electricity supply.

A key role for governments is to improve and support the competitive market framework needed to provide commercial market participants with the highest possible level of transparency. For example, the ability of electricity and gas markets to provide cost-reflective price signals is a prerequisite for balanced responses from market players.

Diversifying Fuel Supply for Power Generation

Fuel supply is the first step in the chain leading to reliable electricity supply. Some fuels, such as coal, natural gas, uranium, biomass and oil, are traded. Supply and demand are cleared through trade that establishes a market price. Volatility in supply and demand is transferred into volatility in prices. These markets may not work ideally, and in themselves rely on functioning supply chains. Other fuels, such as hydro, wind and solar, are essentially free. Their supply is cleared by nature rather than by market, but the volatility of supply also has obvious significant impacts on project profitability.

Most fuels for conventional power generation are traded internationally. Each of these traded markets is complex and has many variations depending on location and quality. Coal and uranium markets can be labelled as global markets. Gas markets are still regional, but are becoming increasingly linked together with liquefied natural gas (LNG). Prices in fuel markets, both historic and expected, constitute an important part of the incentives for commercial market players to diversify. Figure 2.6 shows some of the most important price indicators for natural gas, coal and uranium.

The world's proven natural gas reserves are equal to 64 years of consumption at current rates. Close to 56% of these reserves are found in Russia, Iran and Qatar; reserves in OECD countries represent less than 10% of the world total (IEA, 2006b). Natural gas markets are under pressure for change in several interlinked key areas. Liberalisation is one driver for change. Natural gas markets in United States started liberalising in the 1980s, with the United Kingdom and others following in the 1990s. European Union gas market directives set a framework for EU-wide gas market liberalisation.

Increased use of natural gas for power generation is another driver for change. Natural gas demand for power generation increased strongly over the last 15 years, driven largely by the development of the CCGT technology and the low gas price. During the 1990s, gas demand for power generation increased 6.7% annually; from 1999 to 2003, it increased 3.2% annually, reaching 422 bcm (IEA, 2006d). Such strong growth in natural gas demand has put supply under pressure. North America has large gas reserves, but is forced to take more expensive developments in to use and will see increasing

Figure 2.6

Energy prices are volatile to varying degrees

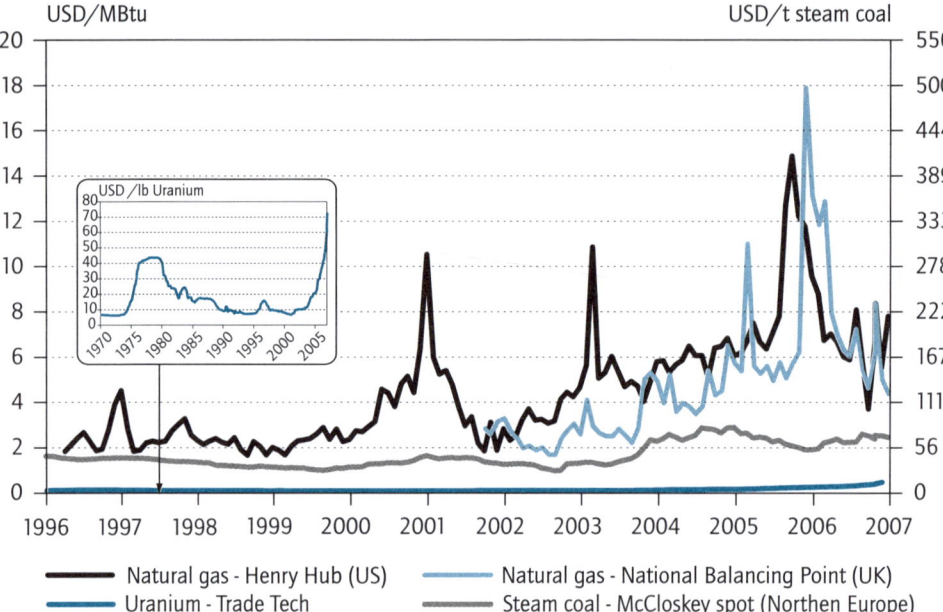

Sources: NYMEX, APX, Trade Tech and McCloskey Coal.

import dependence. Japan and Korea have a very high dependence on imports from non-OECD countries. Yet, the most marked change is in Europe. Natural gas production in IEA Europe has not kept up with the increase in demand. Europe is dependent on imports from non-OECD members corresponding to 40% of demand. This is projected to increase to almost 50% by 2010 and 75% by 2030.

A final important driver for change is the development of trade in LNG. This technology played a significant role for many years in Japan (since 1969) and South Korea (since 1986), but increasing gas prices have made LNG competitive with pipeline gas in other markets. LNG liquefaction and re-gasification plants are now under development in several countries. LNG trade flows are projected to account for about 11% of global gas demand by 2010, about double current volumes. LNG capacities are still too low to ensure full convergence of prices in a truly global gas market. However, increased capacities are expected to continue to act as a converging force.

Uranium resources are abundant. The Nuclear Energy Agency (NEA) has tracked uranium resources and production since 1965 in its annual «Red

Book» publication (NEA, 2006). Uranium resources are categorised by the estimated costs of exploration. Reasonably assured resources (RAR) at exploration costs of USD 80/kg of uranium (kgU) is one category. The volumes of RAR (USD 80/kgU) as a share of actual annual requirements have remained relatively stable since the mid-1980s, reaching 46 years of use in 2003. This reflects the fact that development of resources kept pace with roughly a doubling in annual requirements during the same period. Uranium resources are well distributed across several countries. Figure 2.7 illustrates the country distribution of a broader categorisation of resources, identified conventional resources with exploration costs below USD 130/kgU.

Figure 2.7

Identified conventional resources of uranium, recoverable below USD 130/kgU, are well distributed among several countries

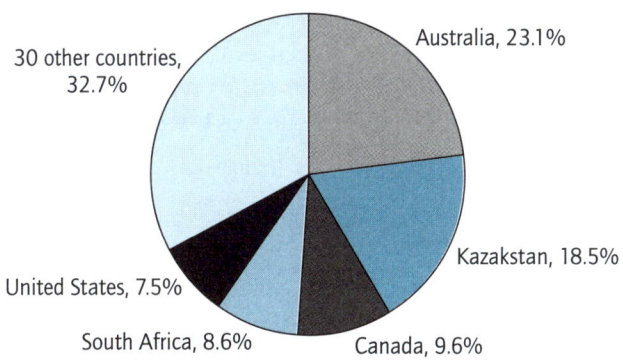

30 other countries, 32.7%

Australia, 23.1%

Kazakstan, 18.5%

Canada, 9.6%

South Africa, 8.6%

United States, 7.5%

Source: NEA, 2006.

Production of uranium (U_3O_8 or yellowcake) is highly influenced by the changes in perceptions of nuclear power and also by the developments in the military use of uranium. Industrial production of uranium commenced in 1945, but mainly for military use. Production for civil use increased rapidly with the development of nuclear power from the 1950s. Successful development of nuclear power during the 1960s and 1970s created expectations for rapid increase in nuclear power and, consequently, led to pressure from demand to build up uranium stocks. This triggered a sharp increase in prices from 1975, which again led to a sharp increase in production capacity from some 50 000 tU/year in the early 1970s to some 80 000 tU/year in the early 1980s. The serious accident in the Three Mile Island nuclear power plant in Pennsylvania (USA, 1979) and the nuclear disaster in Chernobyl (Ukraine, 1986) shifted public opinion against nuclear power. Negative reactions led to cancellations of planned new builds and a less sharp increase in global nuclear power capacity build up. More than 100 plants were cancelled just in the United States. Nominal prices slid rapidly to about half the price of the peak in 1978, and halved again during the next decade.

The end of the Cold War created an important "secondary source" in addition to primary production. Important stockpiles that had been held for military purposes were made available for commercial use. Production capacity declined to some 50 000 tU/year from 1991, about the same level as before the 1975 increase. Worldwide annual production of uranium since 1945 exceeded worldwide annual requirements for civil nuclear power by 20 000 to 40 000 tU/year. This was true until the late 1980s. Annual production fell below annual requirements in 1990, since which time it has been necessary to draw 20 000 to 30 000 tU/year from global stocks. In 2003, stocks (cumulative global production less cumulative global requirements) were still at some 750 000 tU, worth more than 10 years of existing reactor requirements assuming that all are available for commercial use. In practice, an unquantifiable fraction will be reserved for military purposes. Although variable, 10 years is also the expected time it takes to explore and develop a new uranium resource.

The uranium market experienced a marked change in mid-2003. Prices increased from some USD 10/lb U_3O_8 to the current high of USD 70/lb U_3O_8. This price hike resulted from the renewed interest in nuclear power as well as the immediate balance of existing stockpiles, production capacity and requirements; a similar market development as the one observed in the mid-1970s. The uranium market responded. Annual expenditures in uranium exploration increased from a level of some USD 110 million in the late 1980s and early 1990s to some USD 200 million in 2005 (NEA/IAEA, 2005).

Coal is also an abundant fuel relatively evenly dispersed across regional markets. By 2005, proven coal reserves were at 909 billion (bn) tonnes, potentially covering coal consumption at current levels for 164 years. More than 65% of proven reserves are concentrated in four countries (the United States with 27%, Russia with 17%, China with 13% and India with 10%); another 20 countries and regions have substantial shares. Total coal demand was 5.9 bn tonnes in 2005, with 2.2 bn tonnes consumed in China and 1 bn tonnes consumed in the United States, by far the two largest consumers. Coal consumption in China increased by 350% from 1980 to 2004. *WEO 2006* projects that Chinese consumption will increase to 3.9 bn tonnes by 2030, a considerable share of total global projected consumption of 8.9 bn tonnes. Increased demand for coal for power generation, almost entirely in the form of steam coal, represents 81% of the projected increase. (IEA, 2006). Most coal is currently consumed in the country or region in which it is produced. Only about 16% of total hard coal production (coke and steam) is traded between countries. Australia, Indonesia, Russia, South Africa and China are the largest exporters.

Uncertainty and volatility in the supply of fuels, including those driven by both markets and nature, call for diversification of electricity generation options and diversification of fuel suppliers. Well-functioning fuel markets – *i.e.* markets that are responsive to changes in fuel supply and demand at different times and places – reduce the vulnerability of the fuel supply. There is significant scope for improving markets by increasing transparency, pursuing more cost-reflective pricing, enhancing liquidity, and striving for better management of tight situations. Long-term contracts are an important component for producers and consumers for risk management. This is particularly true for gas markets, in that the price of gas is an important driver for diversification for commercial market participants.

Secure supply of fuels for power generation is often discussed in the context of public goods or market failures, referring to the severe financial consequences that would result from a disruption of supply. The resulting situation would be similar to the economic recession caused by the shocks in the oil market in the 1970s. However, secure fuel supply is not a public good in the same sense as, for example, the environment. Commercial market players have no direct incentives to reduce greenhouse gas emissions; the effect from individual action is marginal and everybody would benefit (free ride) from individual action. Government intervention is necessary to "internalise the external costs" of damage to the environment resulting from power generation. Secure fuel supply for power generation is different. An owner of a CCGT would suffer severe, direct private financial losses from loss of gas supply or from market power abuse in the gas market. The point for governments is this: having a narrow focus on operating a profitable business, individual commercial market players may fail to take into account *all* economic consequences of their actions. There is scope for government intervention if, in the event of disruptions in the gas market, for example, total losses to society are larger than the aggregate private losses.

Investors in power generation may not properly or fully account for security of fuel supply when choosing fuel source. Bridging the gap between an abstract understanding of fuel security as a public good and an analysis of the real costs and probabilities is a significant challenge. The scope for market intervention that unambiguously improves the level of fuel diversification is unclear and difficult to assess. A study exploring the prospects for intervening in the Dutch electricity market, with reference to security of fuel supply, concludes that benefits of intervention do not add up to the costs (de Joode *et al.*, 2004).

Evidence suggests that the electricity industry has, until now, been responsive to changes in energy supply markets. Natural gas prices were low until 2003-04. Figure 2.2 illustrates that at gas prices below USD 4/MBtu, CCGTs are very competitive with alternative generation sources. Low gas prices were an important driver for the boost in natural gas-fired power generation in many countries in the 1990s. CCGTs are still built, but now mainly for other beneficial features such as high flexibility, while there is some concern about too high dependence on natural gas in some countries. Higher gas prices are now an equally important driver for renewed interest in coal, nuclear and wind power. In Texas during the past decade, almost all new-builds were CCGTs, the only exception being some wind power in recent years. Looking forward, more than two-thirds (12 of 17 GW) of publicly announced new generation plants are coal-fired, most of them to be commissioned by 2009. Nuclear power is also planned in Texas, to be commissioned by 2014. German market participants have responded by planning for substantial increases in coal capacity. Increasing gas prices are also an important driver for the development of LNG markets, acting as an important source of diversification of gas supplies.

Governments play an important role in that they control the regulatory framework for domestic markets, such natural gas and electricity. However, governments are not in control of international markets or domestic markets of energy exporting countries. Thus, bilateral and multilateral discussions are important to develop mutually beneficial market co-operation. Intensity of international energy talks has increased with the recent increase in oil and gas prices.

Diversification for Secure System Operation

Levelised costs cover all direct costs and may incorporate indirect CO_2 costs, as illustrated in Figure 2.4. These costs are at the bus bar – the connection point to the electricity system. However, there are other important economic factors that determine the economic performance of a generation plant. Several factors will influence the real-time operation of a specific generation plant and the system to which it belongs; in some cases these factors add value to a project, in others they incur costs. The most important factors are operational flexibility, reliability, availability and size of the plant.

Electricity consumption varies on short notice. As electricity cannot be stored economically, the system needs to have access to resources that can respond to these changes with adequate flexibility. Flexibility is required on several different levels – seasonal, daily, hourly, minute-by-minute, and instantaneous. But maintaining flexibility can be costly. Units that are kept spinning as reserves will run for fewer hours and will need higher prices to recover invested capital. Optimal dispatch has to change, with a loss of efficiency for most conventional technologies. Rapid changes will require careful plant operation and may shorten the life of the unit. Automatic, instantaneous response will require additional equipment. The plants best suited to provide these services will expect remuneration and such a cash flow adds to the total economic profitability of the plant. Hydro power, older coal and oil fired units, and CCGTs are particularly well suited for many of these services. Nuclear and wind power are particularly unsuitable.

Load following requires a set of value-adding services, but there must also be resources (reserves) available for "generation following". Plant availability is less than 100%, reflecting the fact that all generation plants have to be taken offline for maintenance, refurbishment and re-fuelling (in the case of nuclear). All plants fail from time to time and are forced to shut down for repairs. This is usually without forewarning, so alternative resources have to be ready to take over immediately. Ensuring that the system has sufficient reserves to cope with the loss of any single generation or transmission unit – the N-1 criterion – is an often used reliability criterion. Hence, the need for reserves increases with the size of the largest unit. As long as most conventional units were about the same size (400 to 600 MW) this was not a big issue. Since nuclear units are significantly larger (up to 1 600 MW) this adds costs to the system that are indirectly attributable to a specific project. At the other end of the scale are small distributed generation units, which potentially reduce system vulnerability to the outage of any one unit.

The low average and very volatile capacity factors for wind power make the indirect costs of this technology particularly important. In fact, in terms of system balancing wind power plants suffer from frequent forced outages, as their availability is not controllable. Costs and operational challenges of integrating – "refining" – wind power into electricity systems is a much debated issue, considering the increased importance of this technology in several countries. This is explored at greater detail in the final section of this chapter.

Nuclear power also has another indirect, but potentially important driver for diversification, at least in terms of the nuclear plant types. If a failure with generic characteristics is discovered, several nuclear power units may have to be closed at the same time due to safety concerns. This happened in Japan in 2002 and, on a smaller scale, in Sweden in 2006. System reliability may be at stake if a large share of generation capacity can suffer from the same failure in terms of a generic shortcoming in design, plant setup, operational practice or safety culture.

Secure system operation has public good elements, and the appropriate diversification to meet all requirements for system operation will rely on some intervention. It is physically impossible to trade fast enough to secure the safe operation of the system. Some resources must be available to react to frequency rather than price. Secure system operation requires "intervention" from an independent system operator, appointed to act in the interests of the public. It is then up to the system operator to clearly define and remunerate the services it needs to operate the system. It is equally important to allocate the costs on a causer-pays basis. Such interventions are explored further in the final section of chapter 3.

Key Message

Diversification of technologies, fuel types and sources is a prerequisite for an efficient and reliable electricity system. Effective markets are important tools to that end, leaving market players with a key role to ensure diversification.

Government policies that encourage generators to pick winners in terms of technologies and fuel types put both efficiency and reliability at stake. Competitive market players respond by diversifying to balance fuel price risks, but only if incentives are right and if governments effectively leave market players an array of technology options. Effective regulation and support of development of new technologies are particularly important. Governments have critical roles to play in the development of competitive natural gas markets, domestically and internationally.

What About Wind Power and Intermittency?.....................

All generation technologies are intermittent. Hydro plants rely on hydro resources that are intermittent. CCGTs, coal- and nuclear-fired power plants fail from time to time. And wind turbines require wind. Wind power is significantly different, due to the level and character of intermittency rather than the intermittency itself.

The main challenges and cost drivers in manageing intermittency of all generation technologies can be grouped in four categories:

- Basic back-up during times when one technology is not available, be it for lack of wind, lack of hydro, planned refurbishment, lack of daylight etc.

- Balancing resources to manage intermittency in the operational phase, in which short-term variations (*e.g.* in wind speeds and demand) must be compensated by alternative resources that are spinning and idle.

- Operational reserves to immediately compensate for the sudden loss of large portions of resources such as a large nuclear or coal unit, a transmission line or sudden loss of large portions of wind power.

- Networks to connect generation with load, the costs of which may vary (*e.g.* depending on distance to load).

The need for back-up to wind power poses a particular challenge in two dimensions. On the energy dimension, installed wind capacity generates considerably less energy on an annual basis per capacity unit than most other technologies. On the capacity dimension, installed wind capacity is more or less uncontrollable; thus, the availability of installed wind capacity cannot be counted on when it is most needed during peak-load.

Regarding the energy dimension, average annual capacity factors vary greatly from technology to technology. The relatively lower utilisation rates or capacity factors for wind power compared to other technologies is reflected in the levelised costs of wind power; the higher the utilisation, the lower the costs.

Average capacity factors of wind power vary greatly by location and year. Precise calculation of average capacity factors requires detailed analysis, particularly when capacity is increasing strongly. Such calculation is beyond this report, but a rough analysis gives some indications. Average capacity factors in Germany, the country with the highest volumes of installed wind capacity, varied between 16% and 23% since volumes became significant in the late 1990s. In recent years average capacity factors were closer to 16-18%. Average capacity factors in other countries with large volumes of wind power were generally higher: 25-28% in Spain and the United States; and 20-25% in Denmark.

Back-up to compensate for the lack of control of wind power – the capacity dimension – can also have considerable impact on the value that can be derived from wind power. In most systems, available capacity during peak-load is the main constraint.[8] In Texas, a system with a summer peak, historically only 2.6% of installed wind power capacity was available during peak-load. In western Denmark, a winter-peaking system, average availability of wind power during peak-load hours since 2000 was 18%, but this reflects an enormous range: between 74% in one year and 5-6% in three years.

The cost of back-up power is difficult to assess. One way is to use modelling to determine the total cost of an electricity system with and without different shares of wind power, and compare total costs. Various academic studies assess basic back-up costs at some USD 4-7/MWh of wind power when wind power shares are at 20% of consumption (NEA/IEA, 2005). Such costs will, in any case, be highly dependent on the particular electricity system and the share of wind power, as well as on the size of the system, interconnectivity with neighbouring systems, and access to natural resources such as hydro and gas.

In a competitive market the back-up costs will take various forms and will develop over time with the adjustment of the generation portfolio. Back-up costs in the energy dimension will either be reflected in the size of the subsidy or in the profitability of the plant. Back-up costs in the capacity dimension will not result in a specific fee or tariff, but will instead be reflected in the wholesale electricity price. When large subsidised shares of wind power are added to a system that already has excess generation capacity, wholesale electricity prices will be pushed downwards. Consequently the need for subsidies to cover the difference between electricity prices and wind power costs will increase – and the total bill for electricity consumers remains unchanged. In contrast, existing electricity generation will lose from having directed capital into assets that are less needed. The generation portfolio and transmission system will change over time, possibly by adding more CCGTs and interconnections to hydro-rich areas. In some circumstances, that may be an efficient development of the system regardless of the increased shares of wind power, at least to a certain extent. In others, it will add costs to the overall system, which will need to be reflected by increasing wholesale electricity prices.

The capacity value of wind power has been discussed intensely since wind power began to account for more significant shares of generation. Considerations regarding the capacity value of a technology are particularly important for systems in which total installed capacity results from a central planning process. In fact, the central planning process may even dictate the

---

8. Hydro-dominated systems, such as the Norwegian one, are exceptions. They are mainly constrained on energy during dry periods.

composition of the generation portfolio. In markets that rely on trade of energy rather than on additional specific valuation of capacity, the capacity value of wind power is less interesting. In such markets, electricity price signals are expected to be strong enough to direct relevant responses by consumers and by investors in both wind power and in alternative generation resources. If wind power is not reliable during periods with peak-load, prices will be extreme during such periods and give strong incentives to appropriate responses. In the US PJM market, generation capacity is priced separately, with wind power being given a capacity value that corresponds to its historical availability during peak-load periods.

Excess capacity and lack of operational flexibility in wind power can create additional costs as is illustrated in the wholesale electricity prices in the western part of Denmark. Denmark West is a small independently operated electricity system that probably has the highest concentration of wind power in the world. It is a price area in the Nordic electricity market. Some 2.3 GW of installed wind capacity corresponds to 32% of total installed capacity. In 2004, wind power accounted for 23% of consumption in the area. Minimum load is some 1.3 GW. Figure 2.8 shows annual average spot prices as traded at the Nordic power exchange, Nord Pool, for Denmark West. Spot prices are cleared on an hourly basis. The Figure illustrates both averages and averages weighted with actual wind power production in each specific hour.

Prices in Denmark West are volatile, with great variations from one hour to the next, depending on the level of supply and demand. Price differences of some USD 25-50/MWh – and often even higher – between the highest and lowest prices during a day are common. Since 2000, Denmark West experienced zero prices on several occasions, particularly during periods with low load. These always coincided with hours marked by significant wind production, but wind was not the only driver. Denmark also has considerable CHP capacity, with heat demand from district heating. This capacity must also run when there is a heat demand, so zero prices were particularly common during cold and windy weekends when electricity demand was low. Nord Pool now tries to adjust market clearing to allow for negative prices. Figure 2.8 shows that a tendency of oversupply during windy periods costs wind power generators some EUR 1-5/MWh (USD 1.2-6/MWh) compared to the average.

The price difference depicted in Figure 2.8 increases with the share of wind power, but other important factors are also involved. The large price difference in 2003 coincides with a serious drought that affected the Nordic electricity system. This resulted in particularly high electricity prices in Norway and Sweden, reflecting the fact that it was particularly costly not to be able to control generated output that year. From 2004, a reform of the tariff system for local CHP plants provided better incentives for them to operate according to the needs of the electricity system, effectively reducing the oversupply in windy periods and thereby reducing the number of hours with zero prices. The number of hours with zero prices were consequently low in 2004 but started to increase again in 2005.

Figure 2.8

Weighted average spot prices for wind power in Western Denmark are lower than average spot prices

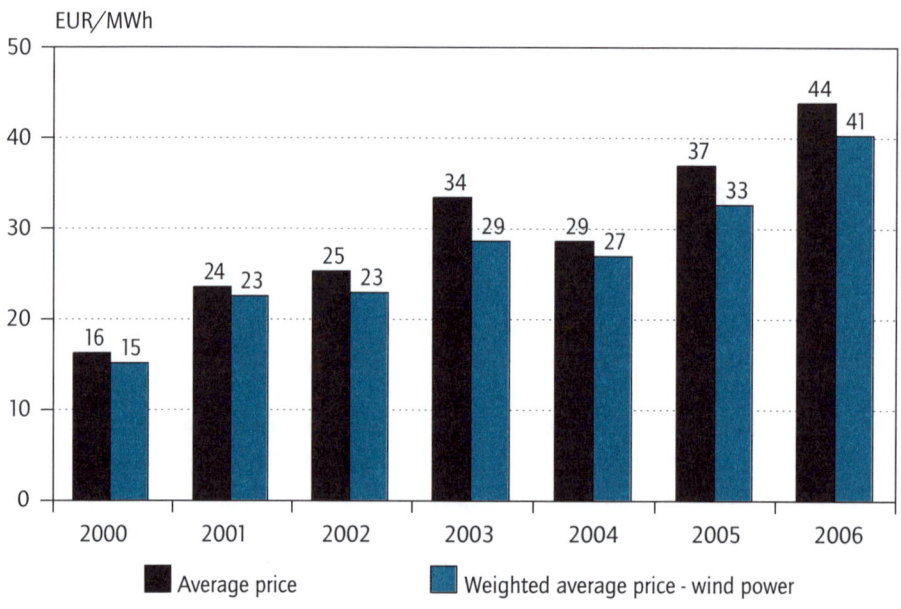

Sources: Nord Pool; and Energinet.dk.

In addition to the somewhat unclear but significant basic back-up costs there are the costs of balancing wind power into the system. Wind power is difficult to predict and behaves differently in different regions. Some places it is more stable than others. Day-ahead spot trade of wind power in Denmark West is based on wind forecasts for the next 13-37 hours. These forecasts have average errors of 30-35% and result in balancing costs of some USD 3/MWh of wind power (NEA/IEA, 2005). An academic study assessed balancing costs of some USD 1.8-2.4/MWh of wind power with wind power shares at 20% in Europe (Auer et al., 2004).

In Denmark West, the normally traded market is the day-ahead spot market. In many other markets it is possible to trade closer to the moment of operation, often up to one to two hours before real-time. There may be scope for lowering balancing costs by allowing wind generators to take advantage of better wind forecasts closer to the moment of operation.

All generation technologies present specific operational challenges and costs, but wind power has additional characteristics that create unique challenges. Specific costs are often difficult to assess and allocate correctly to specific causes. In fact, such challenges are often lowered through diversification and wind power contributes positively when it remains below a certain share of total installed capacity.

The need for operational reserves is one clear example of operational challenges and costs that arise when wind power shares pass a certain level. Electricity systems operate within a set of reliability criteria, which operational reserves are required to fulfil. A main driver for operational reserves is the size of the largest units in a system. Sudden loss of a large unit will require immediate ramp up of resources to replace the loss. The larger the largest unit, the higher is the need for reserves. On-shore wind turbines are small individually (most existing capacity is in the 0.6-2 MW range) and relatively small even when considered as wind farms (on-shore farms are rarely larger than 20-30 turbines; the largest existing off-shore farms are some 200 MW in total). The loss of any one wind turbine or wind farm is not likely to pose any threat to the reliability of the system. In fact, distribution of generation resources to many units rather improves this aspect of reliability.

Wind power becomes a challenge for system reliability when its share increases to relatively high levels. Wind turbines trip not only when there is a failure in the equipment but also when wind speeds reach a certain level, such as a severe storm. Storms are not likely to have the same critical intensity over very large areas simultaneously, but if wind capacity is concentrated in small areas, large shares of generation capacity can be lost within short time intervals. The same effect may result from forecast errors, where a strong wind front is expected to ramp up wind power significantly in a specific hour. If the wind front arrives early or late, the forecast error may be considerable during a short time interval. Denmark West experienced an imbalance of up to 1 800 MW due to wind power forecast errors during a storm in 2005 (Agersbaek, 2006). This is a far larger sudden variation than can occur with the loss of the largest unit in the area, and calls for substantial additional operational reserves. The large share of wind power in Denmark West is only manageable, within a reasonable cost range, because of the strong interconnections with Germany, Norway and Sweden (2 600 MW in total).

A final important cost driver for managing intermittency, including intermittency of wind resources, is the need for networks. Wind power is often connected at lower voltage levels and closer to load centres than other larger conventional units. This saves transmission capacity and grid losses. But if wind power is connected far from load centres or if wind power reaches a concentration level beyond local demand, network costs start to rise with the increase in wind power. Large concentrations of wind power require large transmission capacities to distribute wind power production across larger areas when it is windy – and to import alternative generation when the wind dies down. Hydro power with reservoirs is particularly useful as balancing and back-up for wind power. If hydro and wind are not in the same area it may be cost effective to connect them with transmission, even if this is initially expensive. Studies for the United Kingdom and continental Europe assess the costs of necessary grid extensions to integrate wind power shares of 20% at some USD 3-5/MWh of wind power (NEA/IEA, 2005).

The most important prerequisite for efficient investment responses to integration costs is that investors have the right incentives in the form of cost reflective prices and tariffs. Determining who should pay for integration is complex; in reality, in many existing systems the costs are not properly allocated to those responsible. Ensuring that electricity systems are set up to adapt dynamically and flexibly to changes is a critical feature to support successful integration of wind power. Most electricity systems were traditionally developed to integrate large production units; there were no reason to develop sophisticated controls on lower voltage levels and system operation focused primarily on the few larger production units. Similarly, little attention was devoted to cross-border trade and dynamic co-operation between systems. Large shares of wind power create a need for all these things; at the same time, they add value to electricity systems beyond the ability to integrate wind power. Liberalisation and introduction of effective markets are among the most important necessary developments to enable sufficiently dynamic trade and co-operation.

Effective locational pricing is particularly important to give wind power investors efficient incentives, and to increase transparency of both the need for and the costs of new transmission lines. Wind developments in the northern part of Germany have had significant impacts on trade with neighbouring systems. Available transmission capacity across country borders is often adjusted according to wind forecasts in order to avoid internal bottlenecks, effectively socialising integration costs across consumers in the entire northern part of Europe.

Total costs of integrating wind power are likely to be in the range of USD 5-15/MWh, depending on the location, the characteristics of the electricity system and the total share of wind power. This figure is based on wind power shares in the 15-25% range of total capacity and the existence of a reasonably well connected electricity system. Costs also depend critically on the effectiveness of the electricity market and the network regulation. Poor markets will increase costs and will make costs more opaque, resulting both in inefficiencies and ill-targeted renewables policies. Adding USD 5/MWh or USD 15/MWh to the levelised costs of on-shore wind power may very well have a pivotal impact on the competitiveness of wind power. At USD 59/MWh in the low discount rate case, wind power has lower levelised costs than pulverised coal at a CO_2 price of USD 11/tCO_2 (Figure 2.4). If one adds USD 15 to these costs, the CO_2 price must rise to USD 30/tCO_2 for wind power to stay competitive with pulverised coal. In contrast, a USD 5/MWh integration cost raises the required CO_2 price to some USD 20/tCO_2.

On-shore wind power is still in the final phase of a transition to become a conventional technology. But the large shares of wind power in some countries, and the ambitious plans in many others, create some transitional problems. Efficient, reliable and environmentally responsible electricity systems require that market players have incentives to respond to the immediate needs of the electricity system. When investments in one technology are mainly driven by subsidies and are dislocated from all other relevant decision parameters, overall efficiency and reliability are at stake. When shares of wind power reach 10-20% of installed capacity (as in Denmark, Germany and Spain) subsidy schemes must also include incentives for efficiency. A feed-in tariff, which has driven the development in these three countries, effectively shields investors from many of the incentives for efficiency. Wind power in Denmark and Spain are now integrated into the electricity markets through the provision of a fixed premium in addition to the electricity price. This ensures that the profitability of wind power actually fluctuates with the price of electricity, introducing some basic incentives that are aligned with the overall objective of efficiency.

Renewable support systems based on tradeable obligations are a preferred subsidy scheme in several liberalised electricity markets. Such systems have been introduced in several US states, Australia, Japan, the United Kingdom, Sweden and Italy. Support schemes based on tradeable obligations are dynamic and transparent, and leave many incentives for efficiency with the investor. All support necessary is reflected in the price of certificates. This makes the full costs of a renewable support policy fully transparent. It also allows the system to go through a smooth transition to full market maturity. When the costs of wind power have decreased to fully competitive levels, the certificate price should converge to zero, making the extra support obsolete (*e.g.* beyond a CO_2 pricing scheme). This policy is still relatively new and has had mixed results. Clearly, there are still many lessons to be learnt from pioneering markets about effective implementation of tradeable renewable obligations.

Support systems based on obligations and tradeable certificates are compatible with liberalised markets. They include potential for significant additional efficiency improvements. If renewables are supported with the objective of reducing greenhouse gasses and perhaps improving security of supply, there should be scope for co-operation between states and countries. Presuming that most objectives are global or related to regions rather than countries, cross-border co-operation offer opportunity for improved efficiency. Allowing national renewables objectives and targets to be met in those regions that are richly endowed with natural resources, such as wind and biomass, may improve efficiency considerably. Obligation- and tradeable certificate-based support systems could fulfil such objectives. However, such joint obligation-based systems have not yet materialised anywhere.

Key Message

Competitive markets are an effective tool to integrate wind power at least cost, but it only works if support schemes create appropriate incentive for wind investors.

The relatively higher intermittency of wind power adds costs, which increase with the share of wind power. Effective markets and network regulation are important tools to ensure integration of wind power at least cost. Efficient integration also relies on subsidy schemes that prompt investors to be motivated to contribute towards an efficient and reliable electricity system, rather than focusing only on the development of a specific technology.

RISK MANAGEMENT
IN A NEW INVESTMENT PARADIGM

Across OECD countries, substantial investment is required to replace ageing plants and meet increasing electricity demand. The *World Energy Outlook 2006* (IEA, 2006b), projects that some USD 2.2 trillion is required for new generation capacity. Similar investment cycles arose in the past, with periods of more investment and periods with less (IEA, 2003). In reality, given the size of OECD economies, the amount of investment required is not the primary challenge; there is no reason to doubt that the funds will be available. Rather, the real challenge lies in encourageing investors to direct available funds towards electricity projects. For this, it is necessary to ensure that the rewards in the electricity industry can attract the necessary funds away from competing projects and investment options. This implies establishing a framework that sufficiently rewards invested capital – and that allows for adequate investments in a timely manner, without triggering over-investment. Over the past decade, several companies in various markets have focused investment efforts on mergers and acquisitions. This trend of acquisition of operating assets must now give way to strategies that also add investment in new capacity.

The overarching new development that drives these new trends is the management of uncertainty and risk, which results from the recent introduction of competition. In fact, management of risk and uncertainty in competitive markets is fundamentally changing the traditional investment paradigm.

Traditionally, investments were made within vertically integrated companies according to a planning model[9] and a system of regulatory scrutiny. Once approved, all costs were passed on to rate payers. The efficiency of these decision models, which are still used in many places, depends on the quality of the regulatory approval process and the incentives they create for companies. In a period of steady consumption growth and limited environmental constraints, such processes may be able to deliver relatively efficient outcomes – if they are managed well. However, they have a critical shortcoming in that it is difficult to properly account for real uncertainty and risk in the decision-making process. In the end, consumers pay all costs, including the cost of risk. Linking consumers to generation investments up front through such regulation can reduce the cost of risk, but it also tends to undermine incentives for efficiency. One symptom of this shortcoming was

---

9. For example, the Wien Automatic System Planning package (WASP) for power generation expansion planning, developed by the International Atomic Energy Agency.

– and still can be – a tendency to overbuild. In many countries, it also creates a tendency to politicise the sector and to "pick winners" in terms of preferred generation technologies – different winners in different countries.

With the introduction of competition, risks can no longer automatically be transferred to rate payers. More appropriately, risk remains with those who actually make the investment decision and who are, indeed, the party best able to calculate and account for risks. Incorporating risk into the investment decision has fundamentally transformed the way investments are made. Different technologies have different risk profiles, thus factoring in risk as a real cost element ultimately alters investment choices.

The introduction of competition in the power sector has evolved during a time when uncertainty and risk have increased considerably in many related areas. Significant uncertainty is connected to environmental policy. In some areas, such as pollution control, environmental concerns have already resulted in new standards and requirements. In others, such as climate change, the policy instruments are still far from settled. Demand is also changing character, marked by a slower total demand increase in most OECD countries and by relatively strong growth in peak demand in some countries. All these factors add to the risks and uncertainties that need to be assessed and taken into account in efficient investment decisions. Competition is an effective tool to that end.

Still, liberalisation and the introduction of competition are highly controversial issues. Well-publicised cases of failed or unsteady reforms (such as those in California and Ontario) and the slow progress in many EU countries ignited wide scepticism.[10] Liberalisation also led to significant re-distribution between stakeholders. Those who suffer immediate losses through redistribution are often very outspoken in defence of what they perceive as earned rights. In contrast, it is more difficult to identify the beneficiaries and they are less likely to defend the process in an organised and high profile way. Special measures to protect specific groups of generators or consumers often create barriers to realising the efficiency gains that liberalisation should bring to the total economy. Scepticism on the part of investors and lack of whole-hearted, committed government support put the liberalisation process itself at risk, leaving markets in limbo between a regulated system and an effective competitive market.

---

10. The background and events that led to these outcomes are described thoroughly in academic literature and are also well explored in IEA publications, most prominently in Power Generation Investment in Electricity Markets (IEA, 2003).

This chapter examines the challenge of managing risk in investment. First, it explores the incentives at play in competitive markets, with emphasis on the importance of price signals and the dynamics of investment decisions in a competitive environment. The second section focuses on how liberalisation makes risks more transparent, thereby facilitating more balanced and well-informed risk management. The final section acknowledges the shortcomings of liberalised markets, some of which are due to the complex nature of the product and others that originate in the inherent volatility and risks. The last section focuses on the use of capacity measures as an extra incentive for investment.

Competition Works for Efficient Investments

Competition in liberalised markets leads to optimal investments under ideal conditions (Caramanis, 1982). As with most economic theory, conditions are rarely ideal and assumptions are rarely a perfect reflection of the real world. The merits of various alternative solutions and outcomes are usually measured in shades of grey rather than in black and white. The role of competition to achieve good outcomes is, from a policy point of view, best assessed by comparing alternatives rather than by referring to some theoretical ideal. Some of the relevant questions are: *How are investments developing and what is the outlook compared to an acceptable minimum? How close are we to ideal market conditions and, more importantly, what can be done to improve them?*

First, *what is an acceptable minimum?* With a well-designed, transparent and liquid market in place, competition gives market players incentives to adopt a just-in-time response to the demands of the electricity system. The dynamics of these responses are markedly different from the previous regulated system. What was regarded as the acceptable minimum according to reliability criteria under a regulated system no longer holds true; the threshold of acceptable minimum should be adapted to the new competitive environment. Liberalisation is a process that is creating its own dynamics, particularly as markets slowly mature and become more robust, demonstrating their capacity to direct the electricity system to efficient outcomes over a longer period. It is crucial for governments to monitor the dynamic development of liberalised markets. Governments need to be able to target and time necessary policy responses and – perhaps more importantly – know when to stay away.

Locational Marginal Pricing

How close are we then to ideal market conditions? Locational marginal pricing (LMP) is the electricity spot pricing model that serves as the benchmark for market design – the textbook ideal that should be the target for policy makers. A trading arrangement based on LMP takes all relevant generation and transmission costs appropriately into account and hence supports optimal investments. LMP-based trading arrangements consist of two distinct challenges; many countries have struggled to establish trading arrangements that give effective marginal price signals and many have failed to appropriately integrate the locational aspect of directing investment.

Marginal costs determine prices in a perfectly competitive market in any sector. Markets clear when marginal costs are equal to the marginal benefit from a consumer viewpoint. This balance is referred to as the marginal utility. This basic formula also holds for the electricity market. However, electricity is a tricky product, which makes the actual organisation of trade critical. Market design in well-established electricity markets varies, often due to particular circumstances in particular markets. Experience shows that – even with these variations – it is feasible to establish trading arrangements that effectively order generation plants according to their marginal costs and, thus, enable clearing of supply and demand (IEA, 2005a).

In order to ensure effective marginal pricing, several countries have formally appointed market operators to co-ordinate trade. Variations on how tightly trade is managed and the level of control exercised by the market operator are intensely debated issues. In Australia, spot trade is centred on an obligatory real-time market operated by the independent system operator, NEMMCO, and based on market clearing at the marginal price. This resembles the first trading arrangements in England and Wales – the Pool. The new British Electricity Trading and Transmission Arrangement (BETTA) is perhaps the opposite extreme. It is entirely based on bilateral trading, in which even the trading arrangements for real-time balancing are bilateral deals with the system operator, National Grid. Payments for balancing power are according to the individual bids – "pay-as-bid" or "discriminatory auction" as it is also called. Between these two extremes there are a number of variations on the level of formalised trade, depending on circumstances.

Choice of trading arrangement is often a matter of achieving balance between ensuring sufficient incentives to prompt response (particularly in tight situations) and minimising the potential to abuse market dominance. Transaction costs are also a determining factor. The range of considerations is so broad that no market design emerges as the clear winner. The Australian market seems to effectively provide incentives for response to critical peak-load (Figure 3.4). The UK National Grid recently raised concerns about inadequate incentives during critical situations due to muted price signals in the balancing mechanism, indicating shortcomings to such trading arrangements that are fully based on bilateral trade (National Grid, 2006).

Transparent maginal pricing does not only rely on effective trading arrangements but also on actual system operation. The way in which system operators manage tight situations can influence trade, often by muting price signals. System operators acquire various kinds of reserves. The way these reserves are acquired and activated often mutes crucial price signals that would otherwise reveal a shortage. When reserves are activated during a shortage, it is often without appropriate pricing. In addition, system operators often make out-of-market demands and arrangements with specific generators – arrangements that should have been part of the transparently traded market (Joskow, 2006). Overall, many decisions by market and system operators can result in non-competitive outcomes, as managing system security is a complex task, particularly during supply difficulties.

Creating trading arrangements that lead to cost-reflective marginal prices, which also effectively signals shortage, has proven to be challenging. Some markets, such as Australia, have made significant advances towards achieving this objective. Introducing the "locational" aspect of locational marginal pricing (LMP) has also proven to be both challenging and controversial. The locational aspect aims to reflect the fact that transporting electricity requires considerable resources and that lower-cost power cannot always flow to zones with higher demand. Transmission systems that function like a copper plate – i.e. that cover a large area without bottlenecks – are unlikely to be economical. Transmission costs are affected by consumers' and generators' choices of location, as well as by the subsequent actual consumption and generation. Locational price signals are essential to creating incentives for appropriate siting of new generation capacity.

Transmission lines are still largely regulated businesses. Regulators have the possibility to impose network tariffs that create incentives for location, a practise that has been adopted by regulators in e.g. the United Kingdom and Sweden. Regulation of transmission tariffs is based on cost and revenue calculations and assessments, and is typically established for one or more years. Hence, network tariff schemes are somewhat rigid by nature, which makes it very difficult to properly account for congestion problems that fluctuate or perhaps even shift direction. Locational network tariffs are important instruments for sending locational signals but the true costs of transmitting electricity are more comprehensively taken into account if dynamic locational signals are also included in the trading arrangements. Locational pricing is taken closest to the ideal in north-eastern US markets and in New Zealand, which price every single node in the system, including grid losses. Other markets (e.g. Italy, Norway and Texas[11]) use a more zonal approach, identifying major choke points in the grid to split markets into different areas. The Australian market is also divided into zones, mainly following state borders. Most European markets, including Germany, France, Britain, Spain and Sweden, comprise single markets with zones defined by country borders.

When locational price signals are absent or inadequate, it poses a serious threat to appropriate investment responses in competitive markets. It is evident that some signals (e.g. locational network tariffs or larger zones) are better than no signals. Trading arrangements with very precise locational signals (e.g. nodal pricing) also have drawbacks that must be considered. They probably add complexity. It is also often claimed that they aggravate problems of market concentration. However, claims of aggravating market power problems are unlikely to justify the muting of locational price signals. The level of market concentration is defined and assessed by the number of market players within a specific region, their relative market shares and the possibility to "import" competition. And physical bottlenecks must be addressed through some kind of trading mechanism: locational pricing enables transparent congestion management. In the absence of locational signals, this trade is transferred to a less transparent "side market" and handled through counter-trading. Locational signals do not create physical bottlenecks; rather the price signal merely makes the problem more transparent.

The effect of LMP that alter prices – creating winners and losers on both sides of bottlenecks – seems to be the real underlying controversy in markets lacking locational signals. Locational pricing carries the risk of creating price differences within a country. Opaque or muted locational prices tend to benefit two groups of stakeholders: generators in areas with high concentration of generation and consumers in areas with high concentration of consumption. This implies that other stakeholders are losing, but it seems to be inherently easier to give than to take away – particularly when the parties losing from lack of locational transparency are unaware of their loss. If the losing parties are primarily located in a neighbouring country, maintaining the status quo becomes even that much more the easy outcome. But such an attitude is

---

11. Texas will move to a nodal system in 2009.

not in line with the spirit of internal markets, and lack of corrective action distorts efficient dispatch and investment.

It must be anticipated that lacking or inadequate locational signals are likely to further distort future investment incentives, primarily for two reasons. First, more intermittent resources (*e.g.* wind power and other distributed resources) will need to be integrated and such resources have strict locational constraints (*e.g.* high concentration of wind resources can have serious adverse consequences on system operation) and locational advantages (*e.g.* some distributed resources can be connected to the grid closer to load). Second, it is increasingly difficult to obtain permission to build new transmission lines, which puts an extra constraint on electricity transmission.

Balancing Supply and Demand

Functioning of trading arrangements is described and discussed thoroughly in IEA (2005a), with particular focus on successful experiences in some pioneer markets. As a first step in understanding the expected investment dynamics in such a competitive environment, it is useful to explore the various system demands and resource components that ultimately drive investment decisions.

Looking first at the demand side, meeting peak and extreme peak demand is the main constraint. Need for peak-load resources varies from system to system. Figure 3.1 shows 2005 load duration curves for England and Wales, Sweden and Australia. England and Wales is winter-peaking, as is Sweden, which also has a high share of industrial load (40% compared to 33% in the United Kingdom) and a high level of electrical heating. Australia peaks during the summer due to the high usage of air conditioners and also has a high share of industrial load (45%), which tends to smooth the duration curve.

Svenska Kraftnät, the Swedish system operator, predicted peak-load in 2005 at 26.8 GW with a normal winter and 28.8 GW in a "coldest-in-10-years-winter" scenario (Svenska Kraftnät, 2005). National Grid, the system operator for the United Kingdom, forecast peak-load for England and Wales at 56.7 GW for the winter 2005/06 in a base scenario, 54.5 GW in a low scenario, and 57.6 GW in a high scenario (National Grid, 2004). In England and Wales, the deviation between base cases and high cases is only around 1.5%. Electrical heating (important in Sweden) and cooling (important in Australia) present a special peak-load problem. Svenska Kraftnät's forecast of a "worst case" is 7.5% higher than the normal winter forecast. NEMMCO, the system and market operator in Australia's NEM, makes regional projections; its worst (hottest) case scenarios are between 5.5% and 8.6% higher than the median case (NEMMCO, 2006).

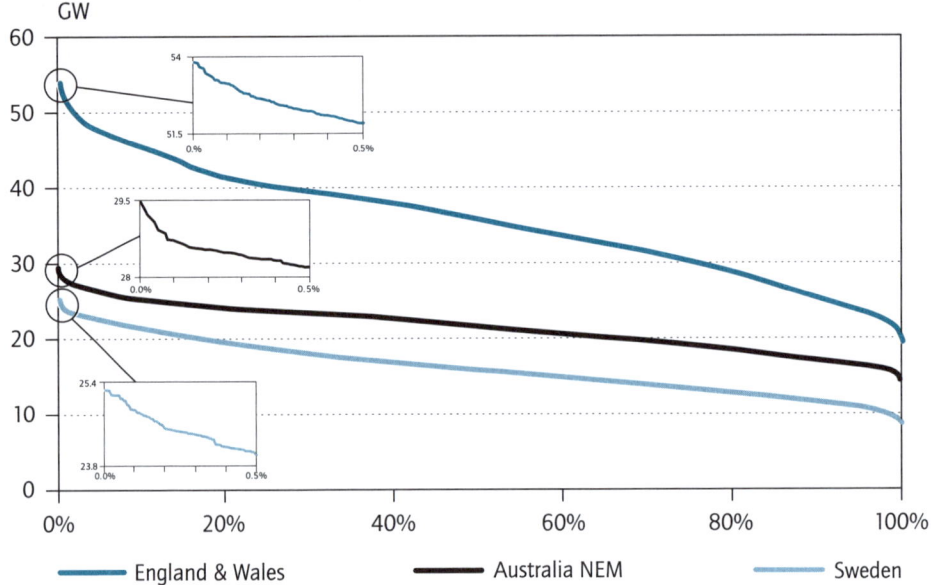

Figure 3.1

**Load duration curves for 2005 show scope for base and peak-load
capacity (Australia and Sweden) and for mid-merit capacity
(England and Wales)**

Sources: NEMMCO; National Grid; and Svenska Kraftnät.

The demand characteristics presented in these examples illustrate some of the difficult tasks an electricity system must manage. The portfolio of generation plants should adapt to match the special demand characteristics in a particular system at least cost, even when risk costs are taken into account. That said, it is obvious that some countries are endowed with particular resources that affect the generation portfolio. Hydro resources most notably change the framework conditions. In systems heavily reliant on hydro power, such as New Zealand (65%), Norway (100%), Austria (70%) and Quebec (95%), availability of energy to meet total demand is often a more critical issue than availability of installed capacity to meet peak-load. Hydro plants are operated to optimise the value of hydro resources over time. Thus, it is rare that all hydro plants in a system will be simultaneously operated at maximum capacity. In fact, average capacity factors for hydro plants in Norway, New Zealand, Austria and Canada are in the range of 40%-60%. Against that background, large variations in peak-load, resulting from high shares of electrical heating in Norway and Sweden (50% hydro), are less challenging than they would be without hydro capacity.

Although hydro resources can serve as an important and flexible capacity resource, most systems are constrained from meeting peak-load by the volume of installed capacity. Available generation capacity in a given area is a finite resource in the timeframe relevant for minute-by-minute system operation. The volume of installed capacity in a competitive market relies on remuneration for investment, which is received through the marginal pricing of electricity output. Markets in which marginal pricing of electricity is the only remuneration are often called "energy-only markets". In fact, "one-price-only" markets is perhaps a more appropriate term, considering that the marginal price is also intended to remunerate invested capital (as is the case in most other product markets).[12]

Invested capital will earn a return during hours in which the price exceeds the marginal costs of a specific plant. Plants with low marginal costs, but perhaps high average costs, will operate in as many hours as possible; this is typical of a traditional base-load plant. Base-load plants earn a return on investment during those hours in which marginal costs of mid-merit and peak-load plants determine the price. In turn, mid-merit plants earn a return on invested capital during peak-load hours.

Peak-load plants meet demand in the few hours in which demand is at its maximum. Peak-load plants also rely on a competitive return on investment to turn a profit. It follows that during peak-load hours, something other than marginal costs of the marginal plant will have to determine the price. In fact, the price is set by the generator with the last available peak-load resource, who can bid this resource into the market at any price – as long as there is no competition from alternatives and no price cap.

One alternative for balancing the system during maximum load is to shed load through forced rolling blackouts. Demand rationed involuntarily through rolling blackouts comes at a very high cost to consumers, which is measured by determining the value of the lost load (VOLL). VOLL has been assessed in numerous studies ranging from USD 5 000 to USD 15 000/MWh, illustrating the great uncertainty in such assessments. In several energy-only markets, a price cap is set at a level based on assessments of VOLL. There are also energy-only markets without price caps at all, such as the United Kingdom, Finland and Denmark.

---

12. *Few, if any, markets are pure energy-only markets. As is explored further in the final section of this chapter, the practical management of reserves creates a grey zone. Energy-only markets are referred to here as markets with no specific remuneration of capacity aiming at a pre-decided minimum level of capacity.*

VOLL can only be estimated and will always be somewhat subjective. However, as a regulatory instrument VOLL serves a purpose towards an efficient electricity sector. They give guidance to the efficient level of investment and the efficient level of safety margins in system operation. If every other aspect of a market (including demand, generation, trade and generation costs) was more or less certain, VOLL pricing might not be the best regulatory tool for providing incentives for investment in peak-load resources. It is not very precise and, thus, will not give very accurate signals. If most factors that are fundamental to the supply and demand balance are certain, adequate generation capacity to meet demand is simply assessed and built. In such a framework, energy-only markets with VOLL price caps may not be the preferred market design. An energy-only market could create a situation in which an electricity system is balancing on a knife's edge – *i.e.* between intolerable shortages with forced rolling blackouts and intolerable market power for the generator with the last resource. In such a world of certainty – the central planner's dream – it may be necessary to let a central system operator decide on the adequate level of generation capacity, in which case a price cap to protect against market power abuse may not be so harmful. However, there are indeed many uncertainties and several alternatives. There are numerous sources of flexibility in the system, important resources and opportunities that risk being lost with a centralised approach to decision making. In such a framework, VOLL is not intended to determine the outcome but merely to set a benchmark.

Flexible Resources

Advanced trading arrangements, improved communication systems and competition have collectively increased the options for supply and improved the tool-box for system balancing. These features have added flexibility through three main sources that offer important potential for the future. First, improved cross-border trade allows for better sharing of resources across larger areas. Second, consumer participation, for example by shifting demand as a response to price, creates a new, cheap and important resource. Third, less traditional resources, such as back-up power and distributed generation, can play new value-adding roles.

Cross-border trade is the first and most important source of increased flexibility. Adequacy of generation capacity is normally assessed within a jurisdiction – a country or region. Clearly defining a relevant jurisdiction is difficult with cross-border trade, and trade across borders is increasing, as illustrated in Figure 3.2. In many countries, cross-border trade intensified with liberalisation also making it more dynamic – possibly shifting direction several times within a day or even within an hour. Integration of larger and larger areas makes it less relevant to link demand and generation capacity at the local level. At the same time, broad integration makes adequacy assessments increasingly diffuse and difficult. However, if trade and system operation are managed well, this is one of the clear benefits of open trade and co-operation. It paves the way to share resources across larger areas and reduces the overall need for investments. Ultimately, demand can be met with a lower margin of generation capacity over peak demand.

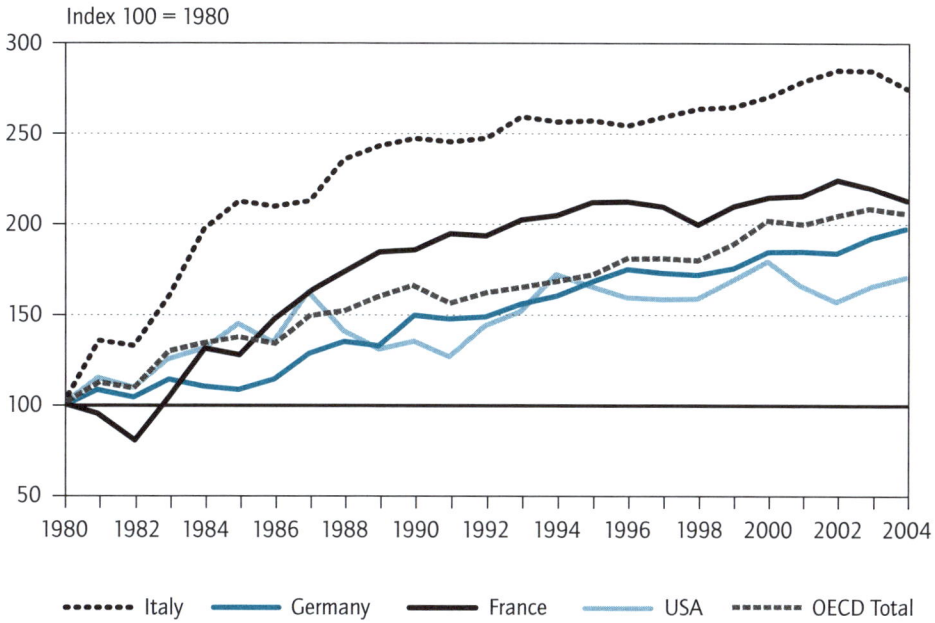

Figure 3.2

Cross-border trade increased substantially since 1980

Source: IEA Statistics.

The second important source of flexibility in liberalised markets is the response by consumers to prices by shifting part of their load to other periods. The flexibility that derives from consumer participation in balancing the electricity system is important, although still developing. It has the potential to become a critical resource in situations of scarcity. The notion of engageing consumers first developed in the 1980s and became an important aspect of demand side management (DSM) initiatives. However, in the absence of competition, vertically integrated utilities do not have clear incentives to opt for demand response resources in a regulated system. At the time DSM was introduced, electricity systems were largely seen to be founded on the two pillars of generation and transmission. Not surprisingly, DSM programmes to engage consumers in balancing the electricity system were rarely very effective. In fact, electric utilities often regarded demand as "price-inelastic" – *i.e.* electricity consumers do not care much about price when making consumption choices and their consumption does not rise or fall when prices decrease or increase. But in times of scarcity, even a very small degree of price elasticity can be enough to deliver the critical resources to balance the system, particularly if prices are allowed to spike. Trading arrangements that effectively establish cost-reflective price signals create the missing link for consumers: the price. Allowing occasional price spikes to reflect scarcity creates the incentive consumers need to respond. Competitive markets ensure that such occasional price spikes, which trigger demand response, cost less than the alternative generation resource.

Electricity demand should be price-elastic in principle. Millions of different electricity consumers use electricity for millions of different purposes. Some consumers can shift some demand easily and cheaply for a short period, but not for longer periods. Some types of demand can shift on short notice, others need longer fore-warning. Some consumers require investment in control equipment, but then become very flexible even on short notice while others already have most of the necessary equipment. Industrial users represent a demand type that fit into all these categories, and much industrial load in competitive markets is already metered in a way that enables these users to participate in the market. To engage smaller consumers, including households, in balancing electricity systems, it will be necessary to equip them appropriately – *i.e.* to provide remotely read interval meters and probably also control and management equipment. Such efforts are already underway, with advanced meters being installed in households in more and more jurisdictions. The largest project to date is a full replacement of 30 million meters in Italy, scheduled to be completed by 2008 (IEA, 2005a). Several countries are also considering a full roll-out of advanced meters. Such meters are one of the very concrete measures that governments and regulators can use to encourage demand response. Costs and benefits of such measures vary from jurisdiction to jurisdiction, and the total benefits will be highly dependent on the ability to create other values such as saving administrative costs and reducing losses from non-payment.

Transaction costs are the main barrier for the potential integration of demand resources. The US Federal Energy Regulatory Commission (FERC) has released a comprehensive assessment of demand response and advanced metering in the United States, including recent research results on the magnitude of price elasticity under various circumstances. The study shows that demand response is important, documenting potentials that correspond to 3-7% of peak-load in most US reliability organisations (FERC, 2006). In the most recent resource adequacy report from UCTE in Europe, demand resources are expected to account for 7 GW during peak-load hours, corresponding to 1-2% of peak-load (UCTE, 2007). Alberta has observed resources able to respond to price change corresponding to 7% of peak-load; system operators in Australia and the United Kingdom report demand response at about 1% of peak-load. In the Nordic market, consumer response to prices reduced demand by some 5 TWh during a drought that hit Norway and Sweden in 2002/03. This corresponded to some 5% of temperature-adjusted demand in Norway; slightly less in Sweden (IEA, 2005a).

Consumers are responding to prices to a certain degree, but they are not yet participating at their full expected potential. The volumes of energy are still not large enough to provide a secure cushion for system reliability. Volumes also need to be larger to provide an effective cap on market power abuse. Two main reasons explain the shortcomings in demand response: small but critical barriers and the overall lack of a real need, at least so far. First, small barriers can play a significant role for most types of consumption. Lack of metering and control equipment thwart demand response, although the largest consumers with the greatest initial potential – the low hanging fruit – usually have the necessary equipment. Trading arrangements and market design still focus largely on the supply side. Retail suppliers need to innovate and offer appropriate contracts that allow for demand response. However, without effective competitive pressure, retailers will not engage in the necessary product development. Second, there is still often a lack of need for demand response. At present, demand response resources are usually only delivered at prices significantly above marginal costs of conventional base- and mid-merit plants. As long as electricity systems have excess capacity, they will never or rarely see tight supply and demand balances. Thus, demand response resources should not be expected to materialise until markets become tight during peak-load. Some markets are still in a situation of over-capacity, as was common in regulated systems. Other markets are only now entering a situation with tighter supply, or expect tighter supply in a near future.

Cross-border trade is the most important source of flexibility and the potential from demand response resources is promising for the future. A third, less conventional source of flexibility comes from the new roles played by small-scale generation. Back-up generation, small-scale CHP, and other distributed resources were traditionally used in specific roles. With competition and liberalisation, they now benefit from access to new markets and can contribute to operational reserves and other ancillary services. In Denmark, smaller distributed CHP units now bid into the market for operational reserves, providing real competition in an otherwise concentrated market. At the same time, the sale of reserves provides important cash flow to these plants. Aggregation of back-up generation, also to serve as reserves, is pursued in projects in several markets.

Flexible resources, such as cross-border trade, demand response resources and distributed generation can improve efficiency considerably and make markets more robust, particularly during tight situations. Figure 3.3 illustrates the main principles in the market-clearing process and the critical impact that flexible resources may have, particularly during times of tightness.

Flexible resources from imports and demand response may be critical in extreme demand situations. The upper half of Figure 3.3 illustrates the principles of market clearing in a system that does not account for flexible resources. Market-clearing prices are determined by the marginal plant at which demand intercepts the supply curve. Some plants, such as wind and CHP, are must-run and will be bid into the market at zero or even negative prices. Hydro power, which is not depicted in the graph, will not enter the merit order according to its marginal costs, which are negligible, but rather according to the expected opportunity cost in any given moment. Nuclear has the lowest marginal costs. Coal is often next, but its order depends on coal and gas prices. Gas-fired plants set the marginal price in many hours. In a few peak hours, the last readily available resources in the system must be used. These might be oil-fired plants and other older plants that were built to operate as base-load but have now been shifted to the end of the merit-order stack. (In fact, it is likely that their owners regularly consider decommissioning and mothballing such capacity.) More rare demand peaks – which may only occur a few hours every year, or perhaps even with several years interval – push the system to it limits and maybe beyond.

The lower half of Figure 3.3 illustrates the importance of flexible resources in the system, particularly during rare extreme situations. Imports, demand response resources and other flexible resources can considerably reduce the market-clearing price; they may even constitute the pivotal resources that actually make the market clear and effectively keep the lights on. Imports can also be transformed into exports, thereby increasing the price and tightening the balance, when resources are needed and are valued higher in a neighbouring market.

Market Experience

Australia is an interesting example of an energy-only market in which prices are allowed to spike during tight situations as a means of triggering

Figure 3.3

Marginal costs determine electricity market-clearing prices.
Imports and demand response add important flexibility to the system

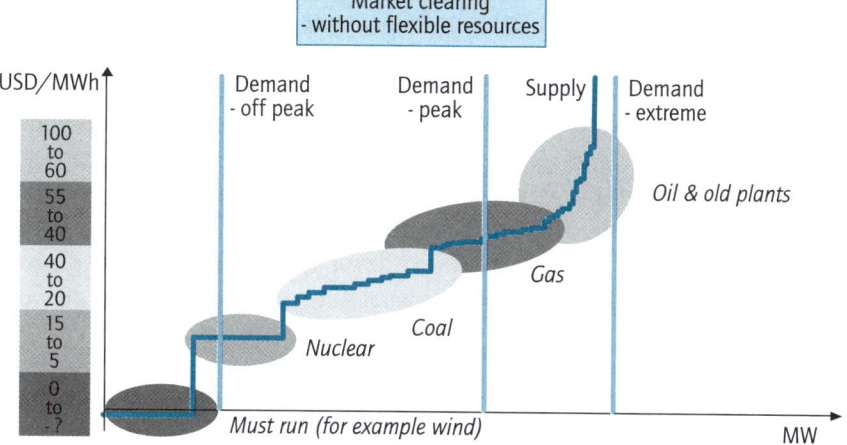

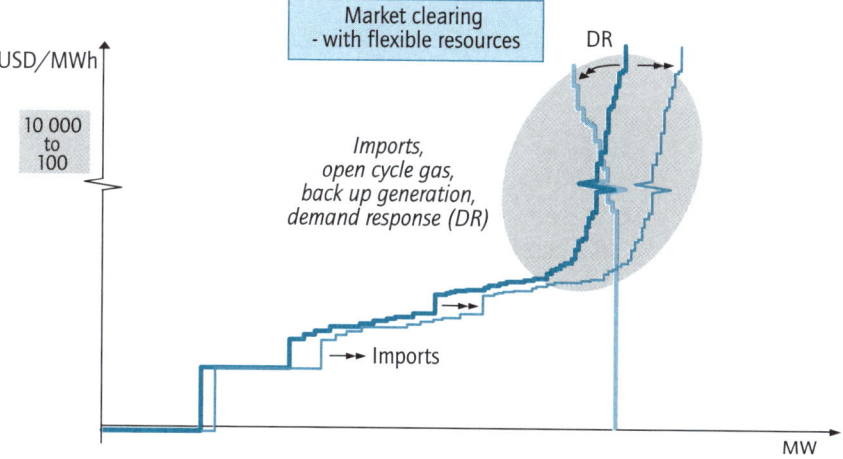

Source: IEA.

investments in new peak-load resources. Figure 3.4 shows electricity prices in the NEM during 0.5% of the time when they were highest in 2005 – this corresponds to 44 hours per year.

There is a price cap in the NEM at USD 8 000/MWh (AUD 10 000/MWh), motivated by an assessment of VOLL. In 2005, prices were above AUD 1000/MWh as follows: fewer than five hours in Victoria; fewer than ten hours in South Australia, Snowy Mountains and Queensland; and fewer than 25 hours in New South Wales. Price spikes have attracted new investment in open-cycle gas turbines (OCGT), a particularly appropriate generation technology for peak-load. System operation was under considerable pressure on a number of occasions and consumers were disrupted. IEA (2005b) thoroughly describes one of these events. But no demand is reported to have been cut off involuntarily due to a shortage of generation capacity. Disruptions have resulted from faults in transmission and distribution systems. Average prices are very low in Australia, compared to wholesale prices in other IEA member countries.

Figure 3.4

***In Australia, wholesale electricity prices spike to extreme levels
in very few hours each year to trigger investment in peak-load resources,
but average prices are low***

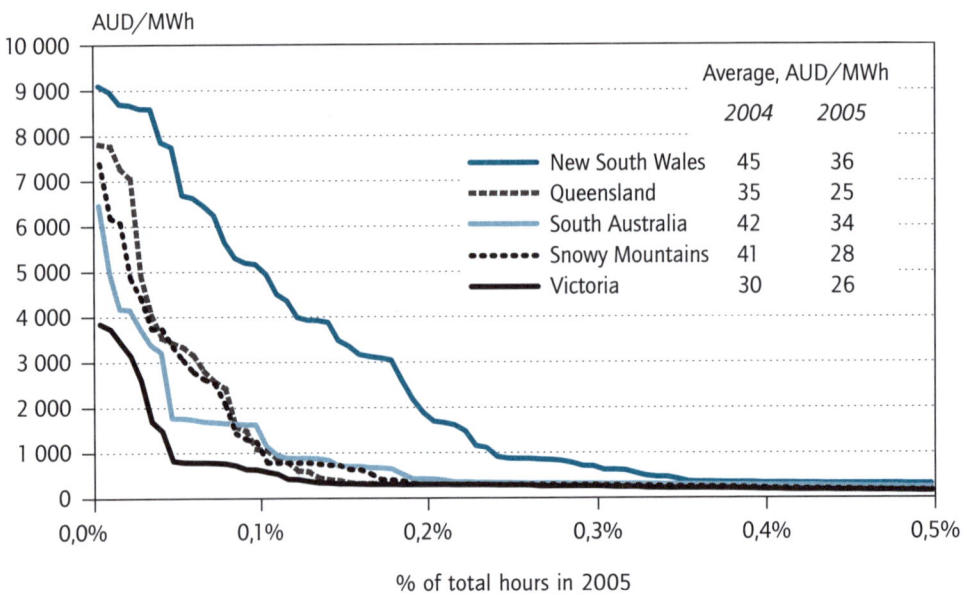

% of total hours in 2005

Source: NEMMCO.

The NEM opened in 1998. By 2004, some 3 700 MW of new gas turbines were installed in Australia, including the states and territories not part of the NEM. OCGTs can be built in as quickly as 6-9 months, so potential market power in the peak-load segment has been short-lived.[13] 2.8 GW gas turbines (OCGT and CCGT) were added in the NEM since market opening, corresponding to about 7% of installed capacity in the NEM. Investment in gas turbines in South Australia corresponded to 30% of installed capacity in South Australia. In 1997, Australia's coal-fired generation capacity was utilised 63% of the time on average. The average capacity factor during the ten years up to market opening (1988-98) was some 60%; during the 1980s it was 50%. By 2004, the capacity factor for coal-fired generation had increased to 76%. In Victoria, the capacity factor for its lignite plants was 89% in 2005 – a noteworthy fact considering that lignite capacity needs to be taken out for regular maintenance. Gas-fired capacity, both in open and combined cycle gas turbines, has enabled increasing utilisation of traditional coal-fired base-load capacity.

Experience from Australia and other markets illustrates that market participants do respond. So far, if trading arrangements allow prices to reflect real costs, responses have been adequate and just-in-time. It is likely that future investment in power generation will be somewhat cyclical, with some periods of excess capacity and low prices, and others with tighter supply and higher prices. Such investment cycles are seen in many other relatively capital-intensive sectors, and have been a common feature of power generation in the past. Competition is, however, likely to smooth investment cycles: decisions will be driven by incentives to use resources more efficiently and to exploit opportunities for trade and co-operation across larger areas. Conversely, if the political and regulatory framework creates too much uncertainty and risk, investment cycles may ultimately become too extreme.

The technological development in power generation is also likely to help smooth investment cycles. Coal, nuclear and hydro plants have traditionally had prominent positions in generation portfolios, typically as base-load technologies. They come in large sizes – and represent large economies of scale – and probably create sharper investment cycles. Low demand growth further increases the risk of building a new large base-load plant. Competition, cross-border trade and the emergence of CCGTs have changed this pattern. In many countries, generation portfolios now also include CCGTs, wind power

---

13. IEA (2003) describes the development of investment in the NEM during its initial phase.

and other forms of distributed technologies. With their smaller sizes, lower investment costs and lower sensitivity to capacity factors (with the exception of wind power), these technologies are less risky in the margins and also contribute to a smoother investment cycle.

Competition creates incentives for just-in-time investments. OCGTs, CCGTs and perhaps also wind power are added quickly when the need is there, effectively adjusting the total generation portfolio to reflect demand in small incremental steps. At one point, gas turbines will set the marginal price in a large enough number of hours to re-establish economic scope for investment in a traditional base-load plant. Investments in traditional base-load plants in liberalised markets are still relatively limited; the nuclear unit in Finland is one unique example. Coal-fired units are now under construction in several countries, most notably in the United States and Germany.

Experiences in Australia and Finland highlight some particular challenges with large size investments in relatively small systems. In Finland, a 1 600 MW nuclear reactor is under construction, which will add almost 10% to installed capacity. Finland is now concerned about capacity adequacy in the short term, until the new nuclear unit is commissioned. In 2002, Australia approved a 600 MW transmission cable between Tasmania and Victoria. Basslink would effectively connect Tasmania to the NEM and would represent more than 7% of installed capacity in Victoria. Construction commenced a year later with the plant expected to go online in the winter of 2005/06. In fact, the plant was commissioned in May 2006; the delay caused a tight supply balance during the winter 2005/2006.[14] Such examples stress the need for careful monitoring and dissemination of information – an issue Australia has addressed through the mandatory *Statement of Opportunities*, published annually by the system and market operator NEMMCO.

Incentives for adequacy and timeliness of investments in liberalised markets are a widely debated question. To date, energy-only markets have not failed to deliver. In well-established energy-only markets, there are no examples of interrupted consumers due to shortages of generation capacity. Substantial investment activity has unfolded in several energy-only markets, as illustrated in Box 3.1. However, lack of competition, lack of effective trading arrangements, lack of clarity on environmental policy and lack of effective regulatory approval processes all raise concerns for adequacy and timeliness of investments.

---

14. *A working group was established in 2005, the Energy Reform Implementation Group (ERIG), to look into the weaknesses that still remain in the Australian gas and electricity markets. One major focus point was investment in power generation and the role and practices of state-owned utilities (mainly in New South Wales and Queensland). A final report of the findings is expected in 2007.*

Box 3.1 . Investments increasing in several energy-only markets

It is often argued that liberalised markets do not create adequate incentives for investment in power generation. This supports the corollary argument that additional government or regulatory intervention is required. So far, outcomes in energy-only markets have not justified these claims, even if a decade of market experience has highlighted several policy issues critical for the investment climate. Figure 3.5 shows development of installed capacity in six different liberalised energy-only markets.

Figure 3.5

Installed capacity increase in energy-only markets, 1990-2005

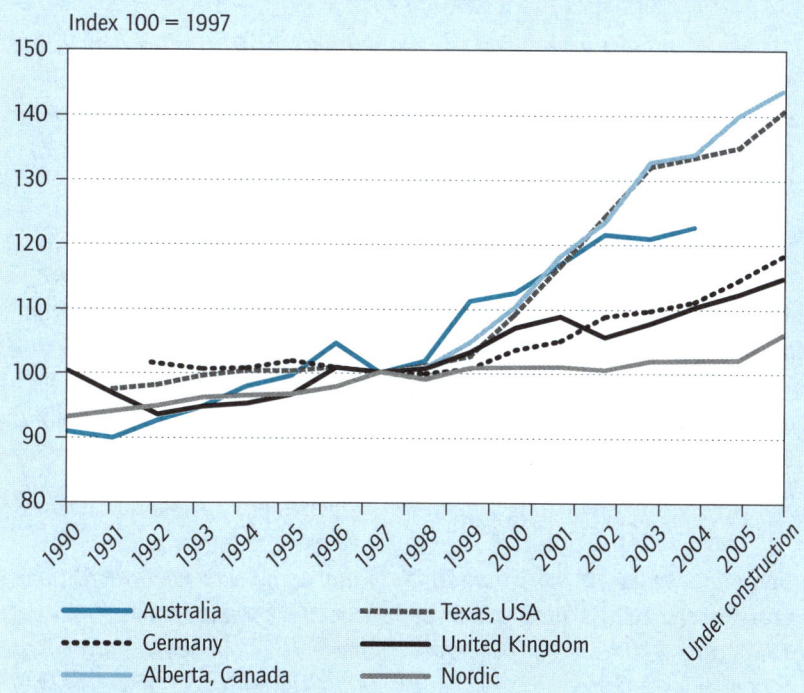

Sources: IEA statistics; EIA; Alberta Government.

The United Kingdom was the first OECD country to liberalise (1990) with privatisation and full market opening in England and Wales. The Pool, the initial trading arrangement, included a capacity payment, which was later abolished with the introduction of the New Electricity

Trading Arrangements (2001). Older coal-fired plants were mothballed and decommissioned from the outset of liberalisation. CCGTs have been added more or less consistently since 1992, both to replace the old plants and to keep up with demand growth. Bigger challenges are looming in the mid and long term (towards 2015-20) as old coal and nuclear power units will close down. Current uncertainty about future environmental constraints and policies, and about public perception of nuclear power, adds substantial risk for investors.

Installed capacity in Australia has increased substantially since the introduction of the NEM in 1998, even though old capacity was mothballed in the process leading up to market opening. Substantial new gas turbine capacity was added to meet peak demand, and average capacity factors of coal plants increased considerably.

Alberta (Canada) introduced competition in 1999. Installed generation capacity and capacity under construction has since increased by more than 40%. Interestingly, Alberta has a price cap at USD 850/MWh (CAD 1 000/MWh), which is normally considered to be substantially below VOLL levels.

Texas (USA) liberalised its market in 2001, which is isolated from neighbouring systems. Installed capacity has since increased by 40%. Most new generation capacity is gas-fired but new coal-fired plants are now planned. Texas has a price cap at USD 1 000/MWh. In 2006, the Public Utility Commission of Texas, the regulator, opted for a capacity adequacy model based on incrementally increasing the price cap to USD 3 000/MWh in 2009, at which time a nodal pricing system will be introduced to create more precise locational signals in the current zonal system (Schubert et. al., 2006).

Germany opened its market in 1998, initially only with negotiated third-party access. Most new generation capacity coming online between 1998 and 2004 was wind power. Some 18 GW of wind power was commissioned between 1995 and 2005. Since 2005, new gas-fired capacity has been commissioned and is under construction.

In the Nordic market, generation capacity was almost stagnating from 1997 to 2005. The Nordic market opened in steps, first with Norway in 1991 and last with the eastern part of Denmark in 2000. Much of the new generation capacity is subsidised wind power in Denmark.

New generation capacity is now under construction, the most prominent project being the 1 600 MW nuclear unit in Finland.

This partial list of recent investment shows strong growth in generation capacity in most markets. However, investment development is not necessarily a good measure for the performance of liberalised markets. Development will – and should – depend on the initial level of capacity, the growth of cross-border trade and the increase in demand. In the end, the best measures are whether lights are kept on and, in the long run, whether costs are lower.

Key Message

Governments urgently need to accelerate the process towards effective and competitive markets with cost-reflective prices that create incentives for efficient use of existing resources and adequately reward new investment.

Competitive markets are proven to serve as an effective tool for efficiency and reliability. But decisive government action and ongoing commitment are required to effectively unbundle networks and system operation, and to establish effective trading arrangements. Half-hearted liberalisation can seriously jeopardise efficiency and reliability; necessary investments may be deferred. Once established, competitive markets must be allowed to function without undue intervention – even when occasional shortages are priced at extreme levels.

Risk Management in Competitive Markets

Uncertainty and its resulting risks are a reality for energy markets. Most of the underlying uncertainties are real and will not disappear, but risk can be managed and shifted between various stakeholders. The management of the risk will, however, have a great influence on the costs of uncertainty. The challenge for public policy is to enable management of risk at the least cost. The key is letting policy makers manage those risks that they manage best, and establishing a framework that allows individual stakeholders to manage risks in areas in which they are the most capable. Table 3.1 lists some of the main risk factors encountered by investors in power generation.

Table 3.1

Main risk factors for investors in power generation

Plant Risk	Market Risk	Regulatory Risk	Policy Risk
Construction costs	Fuel cost	Market design	Environmental standards
Lead time	Demand	Regulation of competition	CO_2 constraints
Operational cost	Competition	Regulation of transmission	Support for specific technologies (renewables, nuclear, CCS)
Availability/ performance	Electricity price	Licensing and approval	Energy efficiency

Uncertainty and Risk in Power Generation

A first step in managing risk is to clearly define and allocate roles and responsibilities amongst stakeholders, including government, regulators, system operators and commercial market players. In some areas uncertainty and risk are best managed from a centralised point of view, on behalf of the well-being of public society. Governments and regulators are responsible for determining environmental standards for generation plants, policies on climate change and procedures for siting of new plants. Through market design and regulation of networks, they also determine the framework within which competing firms will operate. Decisions – or indeed lack of decisions – in these areas creates regulatory uncertainty for investors. This uncertainty carries a significant cost that governments have the power to reduce through clear and credible long-term policy decisions.

But regulatory uncertainty is not reduced through expressed intentions, even though good intentions are an obvious objective for all policy makers. Regulatory uncertainty is reduced credibly through policy action, and sometimes through government inaction, for example by refraining from undue market intervention. Regulatory uncertainty will also reflect current realities, for example the fact that the environmental impact of greenhouse gas emissions is uncertain and public perceptions change over time.

There is also a time factor associated with regulatory credibility. Intentions to remove all regulatory uncertainty with today's decisions are unlikely to support credible risk management by governments for the future. As the world changes and new truths materialise, public policy will have to change with it. Market design and effective regulation are still developing, and important lessons are learnt that should be taken up by pro-active regulators. Conversely, if too much uncertainty is left open without government decisions, investors will be forced to respond in ways that are not sustainable for the environment, for security of supply and for economic prosperity. Considering that reserve margins are declining, and that large shares of ageing capacity will need to be replaced within the next decade, the time has now come for government decisions. Lack of clarity concerning government policy on climate change, particularly in light of the current focus on this issue, contributes greatly to investment uncertainty.

One intention of climate change policy is to drive investment in power generation towards technologies that emit less greenhouse gas. But with too much uncertainty and indecision, there is a risk that investment decisions will be deferred completely, eventually jeopardising reliability. Governments must give investors more firm and long-term directions regarding the future framework for climate change abatement. The ideal is, of course, to achieve global political advancements on greenhouse gas emission abatement beyond 2012, and with a wider scope than is currently reflected in the Kyoto Protocol. Investments must be made in the meantime; in some countries in particular, national climate change policies will have to be established to direct investments in the shorter term. An extension of the EU ETS beyond 2012 would make a significant contribution to reduce regulatory uncertainty in Europe.

Regulatory uncertainty is one aspect of the risks that investors face in the normal course of business. Other key business risks include development in demand, generation technologies, trade, fuel prices and actions by competitors. Demand varies with economic development, as well as with changes in energy efficiency and changes in the composition of demand. Generation technologies are developing to improve performance, to accommodate new needs in the electricity system, and to meet new environmental constraints. Trade is changing with the extension of transmission lines and with the development of new and more effective trading arrangements. Prices of input factors, such as coal, oil, uranium and steel, are changing – as are labour costs. These are all fundamental uncertainties that add to the risks of every investment decision. Such risks are managed best by the commercial market players who make the actual investment decisions.

With the current characteristics of more conventional generation technologies, the business risks an investor faces in many jurisdictions could look like the schematic overview in Table 3.2.

Table 3.2

Qualitative comparison of generating technologies by business risk characteristics

Technology	Unit Size	Lead Time	Capital Cost/ kW	Operating Cost	Fuel Cost	CO_2 Emissions	Regulatory Risk
CCGT	Medium	Short	Low	Low	High	Medium	Low
Coal	Large	Long	High	Medium	Medium	High	High
Nuclear	Very large	Long	High	Medium	Low	Nil	High
Hydro	Very large	Long	Very high	Very low	Nil	Nil	High
Wind	Small	Short	High	Very low	Nil	Nil	Medium

Note: CO_2 emission refers to emissions during combustion/reformation only.
Source: IEA, 2003.

As explored in chapter 2, all technologies are needed in a well-diversified generation portfolio, with the best local mix depending on the specific circumstances of the market. It is also clear that government actions have a great influence on investors' choice of generation technology. If it is not possible to decrease regulatory risks, this may be the pivotal factor for investors to opt for CCGTs to a greater extent than is otherwise preferred.

Uncertainty and the resulting risk will eventually be reflected in the cost of electricity in one way or another. In a regulated system, risks are spread across all rate payers. Some risks may, in effect, be transferred to tax payers, particularly when the sector is dominated by state ownership. In regulated sectors with a great deal of public involvement, unclear separation of responsibilities easily undermines transparency in public policy. For example, costs of climate change policies can become an opaque aspect of the total electricity rate. In a competitive market, the risk is made transparent and can thus be managed through contracts that considerably lower costs. Competition forces investors to take into account all risks when making decisions, rather than just shifting the risks to rate and tax payers. More importantly, risks are allocated to those who are best able to respond to them.

Electricity is inherently volatile and an electricity price intended to reflect all relevant factors must be expected to be equally volatile. The inherent volatility of electricity is re-enforced by the fact that electricity cannot be stored. Price is both the signal and the means of communication across the supply chain in a competitive electricity market. Uncertainties and risks will spill over in the price of electricity, reflecting changes in various aspects of the market. When demand increases, more plants need to operate and more expensive plants set the market price. When the price of coal or gas drops, the price of electricity falls with it. When something unexpected happens in real-time system operation, the price varies rapidly depending on the nature of the incident. When governments constrain the accepted environmental impact, for example by limiting the accepted amount of CO_2 emissions, the cost of operating CO_2 emitting plants increases. In the European Union (EU), an increased price of CO_2 emission permits in the EU Emission Trading Scheme (EU ETS) spills over into the wholesale electricity price.

Uncertainty and risk will, and should, have significant impact on investment choices. If the expected rewards are the same, less risky projects are preferred to those with higher risk. Choice of technology, fuel, timing, location and size will largely depend on basic cost factors in relation to each item. Choice is also linked to uncertainty about future developments and this uncertainty is particularly difficult to assess: the consequences of choice as it pertains to the timing of a project. *What is the "right" time to launch a new project? What is the value of waiting to gain more information?* This uncertainty parameter also has great relevance for public policy. Determining the value of waiting and optionality is receiving greater attention and is addressed in the theory of real options. As it is described and modelled in Box 3.2, the theory of real options can help explain why, for example, CCGTs continue to be one of the preferred options, even when natural gas prices increase.

Box 3.2 . Waiting for governments may prove costly

Companies considering a new power generation investment face risks from many sources, ranging from the very general to the more project-specific. Table 3.1 shows examples of uncertain plant-specific and market variables that companies would typically take into consideration when carrying out a financial appraisal for a proposed project.

Faced with uncertainty in these input parameters, the financial case for a project will also be uncertain. Companies will generally assess the likely range of financial outcomes from their proposed project by modelling several different scenarios. Alternately, they may conduct a stochastic analysis in which different values for the different uncertain

---

variables are chosen in successive runs of a cash-flow model. This makes it possible to calculate an "expected value" for the financial performance of the project.

The expected value can be formally defined as the probability-weighted mean across all the different scenarios or model runs. Companies will use this to assess whether the expected financial performance meets their investment criteria (this methodology is further explored in Box 2.3). However, this would only be a first step in the investment decision. Companies will also take into account the range of outcomes, as these provide an indication of the financial risks arising from the project. Companies will also take account of other strategic considerations such as how the project fits within the company's overall portfolio, how it affects their market share, and whether it gives them entry into a strategically important new market. Real option theory provides one way to quantify the risks associated with a range of financial outcomes.

An important behavioural response to uncertainty is to try to gain additional information to resolve the uncertainty. One way to do this is to wait before making an investment – benefiting from the option value of waiting. If companies have this flexibility, they may ultimately be able to realise a greater project value because they can better tailor their investment decision after watching how events unfold. Conversely, the expected project profitability may be sufficiently high to counter-balance this value of waiting, in which case companies would go ahead despite the future uncertainty. Different technologies have different waiting values. The level of additional profitability required to stimulate immediate investment in the face of an uncertain future can be expressed as a risk premium.

Figure 3.6 provides a schematic overview of how this risk premium can be calculated. It represents the cash-flow for a project that must account for an external event at some future time (T_p), knowing that the event could affect – either positively or negatively – the project's gross margin (the difference between revenue and operating costs).

The option value of waiting creates an additional financial threshold that the project must exceed in order to justify immediate investment. The criteria for investment is therefore no longer that the project exhibits a positive expected net present value (NPV) as in Box 2.3. Rather, that the expected NPV should exceed some minimum threshold set by the risk premium. The greater the range of uncertainty, and the less time

The option value of waiting

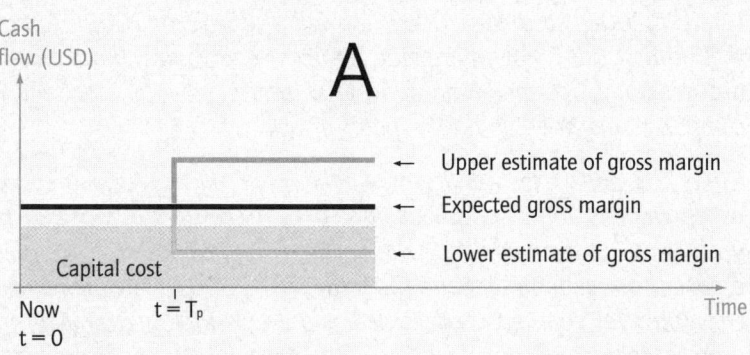

Case A: "Now or never" investment option at t = 0. The company would go ahead with the investment as long as the expected (mean) gross margin is greater than the capital costs, giving a positive value – a positive NPV.

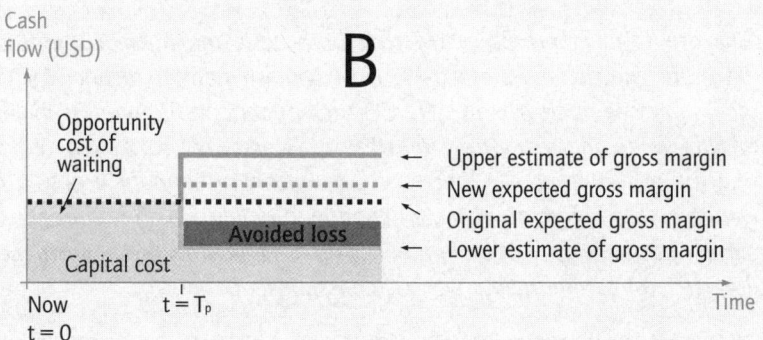

Case B: Wait, and invest after t = T$_p$, the expected time of some event that affects the investment. This flexibility increases the expected gross margin of the project compared to Case A because of the ability to avoid the loss-making situation. The project will only go ahead immediately if the expected project value viewed at time t = 0 increases sufficiently to cover this additional option value of waiting.

Source: IEA.

available until T_p, the greater the option value of waiting will be. Real option approaches have been used quite widely in the literature (Blyth 2006, Edelson, EPRI 1999, Frayer 2001, Ishii 2004, Lambrecht 2003, Laurikka 2006, Reedman 2006, Rothwell 2006, Sekar 2005). Although their formal use is not widespread amongst companies, these principles are generally accepted and provide a useful way to represent how companies might value and respond to risk.

Figure 3.7 shows the results of modelling work that evaluated these risk premiums for the case of fuel price uncertainty and CO_2 price uncertainty. In the modelling exercise, the CO_2 price uncertainty is tied to a policy change ten years into the future (IEA, 2007). The premiums are calculated for coal, gas and nuclear plants, assuming electricity prices are set in a competitive market with marginal prices determined by coal and gas plants. Operating costs for the marginal plant (including fuel and CO_2 costs) are assumed to feed through to electricity prices. Plant cost assumptions are similar to those depicted in table 2.1, but are adjusted so that a cost of USD $25/tCO_2$ equalises the economic case for all three technologies. Fuel price risk is dominated by gas price uncertainty; coal prices are assumed to be relatively stable.

The results in Figure 3.7 show that both coal and nuclear plants can be quite exposed to these risks, even though neither technology uses gas, nor does nuclear emit CO_2. CO_2 risk appears quite modest in these results when the regulatory uncertainty is assumed to be ten years in the future. However, CO_2 price risk can become dominant if only a few years remain before an expected change in policy, which is currently the case – particularly in the European Union – stressing the importance of longer term commitments on CO_2 emission reductions.

In the case of power generation investments, these risk premiums are likely to be recouped through a higher power price. The greater the level of uncertainty and risk, the greater this increase in power prices is likely to be. A recent IEA study on uncertainty estimated that power prices would have to rise by between 5-8% to overcome the risks associated with uncertainty on climate change policy (IEA, 2007). The study also assessed that extending a CO_2 emission reduction period from 5 to 10 years can reduce risk premiums between 4% and 40% depending on the technology.

Technical risks may further increase the required pay-off from projects. In the case of new technologies that have had a limited number of

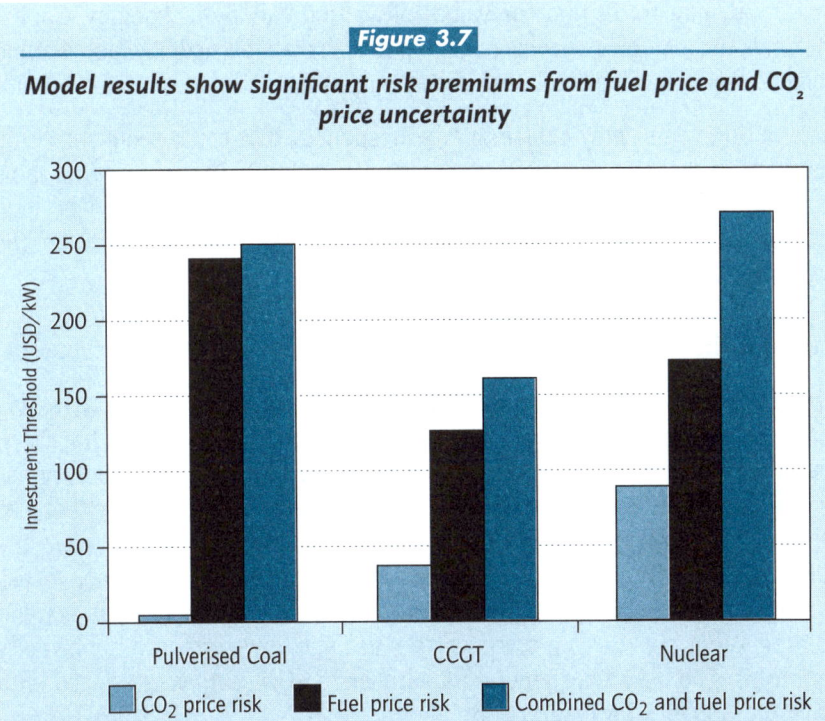

Figure 3.7

Model results show significant risk premiums from fuel price and CO₂ price uncertainty

Source: IEA, 2007.

applications globally, there may be uncertainty about the capital costs or the operating performance. These uncertainties may be resolved as the number of applications of the technology increases. A company may, in this case, have incentive to wait if they can learn from the experience of others. Early movers who take these technical risks with new technologies will, therefore, expect to be compensated through a higher risk premium.

The public policy implication of this is straightforward: the analysis of investment conditions and behaviour in electricity markets needs to properly account for risk, not just for the expected revenue from any given project. Risks can have important consequences for companies' willingness to invest. Compared to a basic financial analysis, such as a levelised cost assessment, the inclusion of risk could indicate an increase in power prices, a tendency to choose technologies that are less risk-exposed, and a possible narrowing of the reserve margin. Creating policy certainty involves ensuring sufficiently long policy time scales.

Box 3.2 demonstrates that the values of waiting vary by technology. Capital intensive technologies, such as nuclear power, have a particularly high waiting value – even if nuclear fuel cost uncertainty is low.

Real option values may explain market responses that could seem surprising considering the cost features presented in chapter 2. The low option value of CCGTs (resulting from low capital costs, short construction times, high modularity and low emissions compared to coal) could contribute to the continued investment in this technology.

Contracting

Risk management trough contracting can lower risks and costs by matching market participants that are willing to share some of the risks. Ideally, it may even be possible to find a partner with a perfectly complementary risk profile: *e.g.* to match a 100 MW generation plant with a retailer that has a 100 MW demand for the lifetime of the plant, both of which have the same appetite for risk. Such a perfect match is unlikely. Thus, active risk management is more a matter of putting together a portfolio of assets and contracts that allow a firm to cover risks but still leave open an opportunity for profits. The risks that generators, retailers and consumers are faced with drive the appetite for contracting.

One strategy for risk management is to vertically integrate generation and retail supply. In a market with retail competition, some consumers are likely to switch supplier if competitors can offer better contracts, but if most consumers will stay regardless of price – "sticky customers" in markets where transaction costs are too high – vertical integration can be an effective "physical hedge". Experience with customer switching in retail markets varies greatly. Large consumers have switched in large numbers in most markets. In contrast, there are relatively few examples of markets in which large numbers of small consumers (including households) have switched. The United Kingdom and Norway are amongst a handful of markets in which small consumer switching is evident (IEA, 2005a).

An alternative strategy is to manage risk through contracts between independent players in the wholesale market. In most markets, utilities have mixed strategies and contracting has increased with the market maturity. Relatively small firms with only generation assets are common in several markets, but it has been difficult for retail companies without generation assets to survive anywhere. Independent newcomers in retail markets would considerably improve retail competition. Thus, it is important that regulators scrutinise the design and functioning of markets in order to remove any undue barriers for independent firms, particularly on the retail side. A liquid and deep market for financial contracts would significantly enhance the ability of independent firms to operate. Such liquidity is only likely to develop on the back of a well-designed and competitive market.

Ultimately, the competitive pressure on incumbent firms will come from new generators – and from the risk of losing customers. Under these conditions, vertical integration cannot stand alone as a risk management strategy in a competitive market. It is too rigid to allow for efficient adaptation to constantly changing market fundamentals. Demand is changing within the hour, day, week, month and year. Retailers win and lose customers every day. Changes in fuel prices change the cost of generating electricity every day. Plant failures may force generation capacity offline on very short notice. The need for refurbishment forces plants offline at certain intervals. Risk profiles change dynamically, thus the basis for a contract decision one day may have changed by the next. If it is easy and cheap to sell and buy contracts, risks will be managed very tightly; contracts worth several orders of magnitude higher than underlying generation and demand may change hands every day.

Allocation of risks amongst various stakeholders in the electricity sector resembles the allocation of risk in many other markets. Inherent volatility in electricity is very high, but the challenges and solutions for risk management are parallel with commodity markets and many other markets. Risks can be shifted between stakeholders through contracts; producers and consumers have risk profiles with opposite signs, allowing for cost reductions. Both of these stakeholder groups have a certain interest in limiting each others' risks – even if their respective risk profiles are different in character. It is likely that most stakeholders will have to accept a premium for covering risks. A consumer may be able to negotiate a long-term contract matching a particular risk profile, but this will probably be at a price that includes a premium on top of the current market price. A generator may be able to negotiate a long-term contract for the entire economic lifetime of a plant, but it is likely to be at a premium deducted from the current market price. Such an arrangement could, in effect, undermine profitability of the project.

Transaction costs are minimised when products are standardised and trade is well organised. This often takes place on a formal exchange or through mediation by brokers in a so called "over-the-counter" (OTC) market. Exchanges and electronic OTC trading platforms have developed in several electricity markets. In the Nordic market, total aggregate volumes of traded contracts increased constantly to 2002/2003. At that point, liquidity dropped in connection with a severe drought, but has since increased again. Liquidity in many other markets was greatly affected by the collapse of Enron in 2001. Enron was deeply involved in energy trading and its financial default had a great impact on traded markets. One particularly significant impact was a change in the perception of credit risk: The collapse of Enron and other large US utilities emphasized the risk of default on a contract. Clearing of contracts through standardised procedures has since become a critical prerequisite for liquid contract markets with low transaction costs. Clearing is the process by which a clearing house or exchange take all the credit risk and it is developing in most markets.

Liquidity is measured in several relevant ways. The critical issue for market participants is whether it is possible to trade at low cost – even in case of an emergency. *What is the spread between the best buy and sell price? What are the fees for trading? How many market participants are registered and active in the market? How many bids are announced in the market in average?* The answers to these questions affect the volumes that can be expected to be traded at any given moment without affecting the price. Actual traded volumes are a good indicator of the overall level of activity, but they are not the only relevant one. Table 3.3 lists liquidity for a number of markets, in volumes traded as a share of consumption.

Activity in traded electricity has developed significantly during the past years. Substantial volumes are now traded in several markets across Europe. At present, information about OTC trading is not very transparent. However, a recent sector inquiry by the European Commission Directorate General for Competition – an important source of information – found that OTC forward trading in Germany and the Netherlands corresponds to 5-6 times the consumption in these countries (EC, 2006). The sector inquiry also reports significant OTC spot trading in Europe: some 5-6% of total consumption in Germany and the Netherlands, and 9% in the United Kingdom. Overall, Nord Pool, the Nordic electricity exchange, trades or clears 5-6 times the total Nordic consumption. Significant growth in trade has occurred in all the markets included in Table 3.3, except in the UK. Exchange-traded volumes are larger than in 2004; in several markets the increase continued in 2006. In Australia, trade on *d-cypha*/Sydney Futures Exchange in the third quarter of 2006 corresponds to the almost entire volume in 2005 – and in March 2007 trade corresponded to 240% of total load in the NEM. Trade on Powernext in France, in terms of shares of consumption, increased by more than 50% in the first half of 2006 compared to the previous year.

Table 3.3

Liquidity in selected electricity markets, 2005

	Traded volumes, TWh			Percentage (%) of consumption			Registered participants (active[1])
	Spot	Forward markets Exchange	OTC	Spot	Forward markets Exchange	OTC	
Alberta[6]	130			263			244
Australia (NEMMCO)	176	34[3]	199	100	19	113	102
France (Powernext)	20	62	355[1]	4	14	79	52 (26)
Germany (EEX)	86	240	3 140[1]	15	43	565	151 (52)
Netherlands	16	52[4]	550[1]	15	48	509	(28)
Nordic (Nord Pool)	176	786	1 316[2]	45	200	334	330 (60)
Italy (GME)	203			63			91
Spain (OMEL)	223			89			544
United States (PJM)	276	2 853[5]		40	417		430
United Kingdom	9		518[1]	2		146	(42)

Notes:
1 EC Sector Inquiry (European Commission, 2006), June 2004 – May 2005.
2 Cleared at Nord Pool.
3 d-cypha trade traded at Sydney Futures Exchange.
4 Endex.
5 FERC (EQR).
6 2004 data.

Sources: AFMA, European Commission, Nord Pool, EEX, OMEL, GME, APX, NEMMCO, Endex, Powernext, FERC, Thon (2005).

US trade data reported in Table 3.3 is of a different character. It is based on the FERC Electronic Quarterly Report, a publication resulting from a mandate, after the Enron collapse, to report all trades between market participants to FERC. Reported trades in PJM increased by 53% from 2004 to 2005. Some of the trade is organised by the New York Mercantile Exchange (NYMEX) and Intercontinental Exchange (ICE). Trade organised by ICE corresponded to about the double of total consumption in PJM in 2006.

The types of contracts vary between markets, depending on the particular needs. All markets trade contracts for the coming month and quarter; contracts further into the future are also common. In 2006, Nord Pool launched trade in contracts for delivery five years ahead of time.

Market players with generation plants and retail customers are obvious participants in electricity trading. Information and market analysis by traders are equally important to optimise the electricity system and achieve all possible efficiency gains; the insight they provide is monetised as a value-added product. Generators and retailers also act as market traders in that sense. In fact, many utilities organise their business so that retail, generation and trade act as independent units to a certain extent. Other players enter the market only as traders, participating without any physical presence in the form of retail customers or generation assets. The presence of pure traders in an electricity market indicates that transaction costs and entry barriers are low. These traders perform an important role in manageing information, matching risks and adding competitive pressure. Pure traders have been present in the Nordic market for several years without interruption. The collapse of Enron in 2001 prompted some purely financial market participants to withdraw from some markets, but they are now returning. More and more banks and other types of financial companies are now engageing in electricity trade. This is also likely to further boost liquidity.

Strategies to vertically integrate generation and retail are being pursued in most IEA countries. Combining a physical hedge of a generation portfolio with a retail portfolio has some attractions, even if it is less dynamic than wholesale contracts. "Sticky customers" create a level of certainty for the generator. Such an approach may be a natural part of a risk-hedging strategy. However, it can also be a threat to competition for two reasons. First, because it is founded on the ability to - uncontested - pass on costs to consumers. Second, this could lead to a vicious circle that drains liquidity: Vertical integration drains liquidity in traded markets, which increases the risk of relying on a traded market, which, in turn, increases the advantages of vertical integration, *etc.* Policy makers and regulators hold two critical keys to ensure that the possible need for some physical hedge does not turn into a threat for competition. One key is to design a marketplace that facilitates smooth and seamless trade, with minimum of transaction costs. The second key is to establish systems that support quick and easy shifting of retail supplier. A certain degree of standardisation of contracts is one aspect of such regulation, but it should not undermine the possibility for consumers to choose to be bound by longer contracts. Longer contracts create hedging benefits that can be shared amongst market players.

Box 3.3 presents an interesting case study of risk management, contracting and financing of a very large generation unit; the 1 600 MW nuclear unit in Finland. The example shows how an advanced mesh of contracts can spread risks across many stakeholders. It also illustrates the important role effective and competitive markets may play in facilitating large investments by smaller and independent market players.

Box 3.3 . Creative risk management to finance nuclear power in Finland

In December 2003, Teollisuuden Voima Oy (TVO) decided to build a new 1 600 MW nuclear unit. Construction commenced in 2004 at the site of two other TVO nuclear units, Olkiluoto 1 and 2. The new unit is a third-generation European pressurised water reactor (EPR) and is being delivered as a turnkey project by a consortium of Areva and Siemens. These vendors carry, to a large extent, the risks of project delays and budget overruns. Total project costs are estimated to be around USD 3.5 billion. OL3 was initially scheduled to be commissioned in 2009 but is now delayed in construction to 2010.

TVO is owned by several Finnish companies. Pohjolan Voima Oy (PVO) is the largest shareholder with 60.2% of the OL3 shares. A majority of PVO is, in turn, owned by various companies in the Finnish pulp and paper industry; the remaining shares are owned by municipalities and municipally owned local utilities. Fortum, a partly (51.7%) state-owned utility, owns 25% of the OL3 shares in TVO. Another 8.1% of OL3 shares are owned by Oy Mankala AB, a fully owned subsidiary of Helsinki Energy (a utility owned by the city of Helsinki). The remaining OL3 shares are with EPV (6.6%), a regional energy procurement company owned by 21 local utilities, which are mainly municipally owned. EPV also owns 8% of PVO. All in all, a majority of TVO-OL3 is privately owned, with a large share of state and municipal ownership.

The project is financed on the balance sheet of TVO, which implies that recourse on loans is not limited to the OL3 project but tied to TVO as a company. This has allowed for 75% debt financing of the project. TVO shareholders injected sub-ordinated debt and equity corresponding to 25% of the finance requirement.

TVO sells its output at cost to its shareholders. This is the key allowing for the high level of debt financing. The OL3 project is covered by long-term contracts that effectively pass all risks on to the shareholders. Thus, risks are spread across the underlying meshed ownership structures. This does not eliminate the real risks. The large consumers and utilities receive generated electricity at cost. If these costs cannot compete with the wholesale price of electricity in the Nordic market, the project shareholders will incur a loss compared to the alternative of buying electricity in the market or producing electricity with a more competitive technology. But the Nordic market also offers a relatively liquid financial market, which creates an opportunity for the final owners of OL3 to manage remaining risks – at least to a certain extent.

Key Message

Regulatory uncertainty affects, delays and deters investment decisions. Government and regulatory action is needed to reduce regulatory uncertainty.

Governments must establish a market framework in which those making investment decisions are also those who carry the real risks and are awarded accordingly. A competitive market framework will allow market participants to manage risks through contracts, which also offers consumers an effective protection against price volatility.

Capacity Measures are the Last Resort ....

Well-designed markets and effective competition have, so far, proven to contribute critically to provide incentives for investments. However, there are limitations to the feasibility of implementing textbook electricity market design in real-world electricity systems. Some of these limitations relate to secure operation of the electricity system. Others relate to possible shortcomings that are often politically driven and may have an impact on the transition to more robust markets.

Electricity supplied in real time resembles a public good. A system operator is required to balance supply and demand in real time, compensating for the fact that electricity cannot be stored economically. In the case of a full-scale black-out, all consumers are disconnected: paying more will not reduce the probability of disconnection, and paying less will not increase the probability. Public goods are often thought to be supplied most efficiently through a system of centralised public decision making. In contrast, liberalisation in the electricity sector is intended to improve efficiency and reliability through the introduction of competition and decentralised, commercial decision making. This paradox between the basic characteristics of electricity and the objectives of competition presents many of the fundamental challenges in electricity market liberalisation.

Competition in this sector is expected to deliver efficient and reliable outcomes only to the extent that electricity is defined as a normal, private, tradeable product. Effective and precise allocation of responsibilities amongst governments, regulators, system operators and commercial market players is a prerequisite for successful liberalisation. That said, it is evident that reducing the responsibility of commercial market players through regulation also diminishes scope for harvesting efficiency gains through competition. Regulation and market design imply a balancing act of increasing responsibility for commercial market players to the greatest extent possible without jeopardising system security. Ultimately, the aim is to tailor risk management of the electricity system such that independent system operators are responsible only for the areas in which they are, indeed, best positioned to ensure the desired outcomes.

Researchers, policy makers and industry have discussed, for more than two decades, the development of market design to successfully introduce competition in this traditionally public sector. Many detailed issues remain unanswered and untested, but there is now sufficient knowledge and experience to understand the key building blocks in successful liberalisation. Many remaining controversies, particularly related to investment adequacy, originate from tradition and from political reality rather than from fundamental uncertainties about efficient market design. The big unresolved question is this: *Will the way electricity systems are traditionally operated and the political environment allow truly cost-reflective prices – even when resources are scarce?* Demand response to price would resolve many issues, so another critical question is whether the lack of price elasticity is an inherent market failure or a transitional problem.

The issue of "stranded assets" has had a significant influence on market design in many regions. Because assets were built and financed under regulatory regimes that offered protection, their actual return on investment is uncertain following the introduction of competition. Markets that were previously mainly state owned, such as the United Kingdom and Australia, could define market conditions before privatising, allowing the consequences of competition to be reflected directly in the valuation of the companies. When privately-owned utilities dominated the scene, as was the case in, for example, Spain, the United States, Japan and Germany, there was less room to manoeuvre in setting up markets. Privately-owned utilities were concerned about receiving a return on investments incurred under a regulated scheme. This was aggravated in the US markets in which neighbouring markets still remained regulated. Regulated utilities received a regulated remuneration for their investments but could, at the same time, compete in neighbouring liberalised markets. Such conditions contributed greatly to the call for special capacity measures as a guarantee of a return on existing assets.

Capacity measures implemented to protect existing assets were deemed necessary to secure support from utilities for the liberalisation process. The resulting risk is that the benefits from liberalisation may be shifted away from consumers, which is currently jeopardising the long-term support for liberalisation in several markets. A half-hearted or interrupted liberalisation can lead to consumers having to pay twice for existing power plants, without ever realising any benefits from competition.

Capacity measures were not the only instruments used to recover stranded assets. When the market opened in Denmark in 1999, the large incumbent utilities complained about their situation and financial prospects in a liberalised market. They managed to convince a political majority that they needed more than USD 1.2 billion in support, paid by consumers over ten years (Folketinget, 2003). At the time, prices were very low in the Nordic market. Prices increased a few years later, and have since stayed on a higher level, due to a generally tighter market. Utilities' profits have been higher since 2002. In Texas, an incumbent was allowed by the regulator to recover losses from a sale of a group of generation units during the shift from regulation to competition – units that later turned out to be very profitable for their new owners when electricity prices started to increase with the tightening of the supply and demand balance. Consumers in Texas are now voicing scepticism over the effectiveness of liberalisation to bring down costs (New York Times, 2006). When it liberalised in 1998, Spain applied a scheme known as the costs of transition to competition (CTC) system to compensate incumbent utilities for stranded assets. This system was relatively transparent, particularly when compared to other *ad hoc* measures. Nevertheless, the CTC system is broadly agreed to have seriously distorted the Spanish electricity market (IEA, 2005c). In several EU countries, the Electricity Market Directives were implemented slowly and with great hesitation. This may also be regarded as an attempt to reward incumbent utilities a protected market position during the transition towards full competition – a grace period.

The following sections explore first the features in electricity supply that do indeed call for centrally co-ordinated action – one being the need for operational reserves. It then goes on to examine the need for additional capacity measures beyond operational reserves. The final part analyses the grey zones between these two concepts, and highlights the scope for softer interventions to manage the transition towards robust competitive markets.

Operational Reserves

Various features and complexities in electricity system operation create particular constraints and demands on generation capacity, in order to uphold acceptable system security.

Resources must be ready to adapt immediately to changes in the system; changes in demand, generation or transmission (demand and generation outside the area). Hydro reservoirs constitute a more immediate storage of energy than gas, coal or uranium, but the distance between a hydro reservoir and an electrical outlet is still too vast to ensure supply without co-ordination by a centralised system operator. The short time period in which a system must reliably respond to changes necessitates the allocation of at least a minimum level of responsibility to the system operator.

Commercial market players are in a position, and have the means, to respond to changes in demand and generation capacity within an annual, seasonal, daily and hourly timeframe. Enabling the right response at the right time requires that the framework in which these players operate offers the key features of competitive markets, including high transparency and effective trading arrangements. Closer to the moment of operation, individual commercial market players lose the overview necessary to enable efficient system balancing: the system operator must assume more responsibility. System operators assume operational responsibility at gate closure, the point in time at which commercial market players are no longer allowed to adjust schedules. In some markets, including Australia, PJM and the UK market, gate closure is 1-2 hours before the moment of operation. In several other markets, particularly in Europe, gate closure is one day ahead of operation. Trade closer to the moment of operation (intra-day trade) is now under consideration in several European markets.

An important part of official system responsibility is to manage operational risks on behalf of the public – *i.e.* to ensure secure operation of the electricity system. In most regulatory and legal frameworks, this responsibility is limited to the short timeframe of actual operation, even if the parts of the supply chain preceding actual operation may put constraints on secure system operation. Events with low probability and high impact will occur. Electricity systems will break down from time to time. The issue for secure system operation is to clearly define the acceptable level of probability.

The probability of blackouts may be lowered to a certain extent, through effective implementation of rules and standards for co-ordinated system operation, but generally the necessary increased security levels come at a cost. Regulators must establish the framework in which system operators carry out their role. This includes a definition of the acceptable system security (measured in lost energy per year) based on the assessed VOLL. One rule served as a standard element of the reliability requirements during the past 20-30 years: The *N-1* criterion, which implies that the system must be able to operate – unaffected – with the loss of any one generation or transmission unit, including the largest. After a certain period of time (usually 15 minutes), the system must be able to handle another contingency event without being affected. New reliability standards are under development in many IEA member countries, most of which are being adapted to reflect the lessons learnt from the large blackouts in North America and Europe in 2003. Some countries are taking steps to apply more advanced metering, control and modelling capacity, including principles based on probabilistic system analysis. Collectively, these changes are affecting the use of the *N-1* criterion (see a thorough description in IEA, 2005b).

In order to fulfil its responsibilities, the system operator must have access to resources or reserves. The least-cost option to secure these reserves usually involves establishing contracts for the necessary resources out of the total pool of assets made available by commercial market participants. Thus, contracting arrangements and incentives for various kinds of reserves will also have a direct effect on the incentives for investment in generation and demand response resources. Payment for reserves is part of the possible revenue stream to power generation assets.

System operators require various types of reserves, as well as other so-called ancillary services, including two categories of ancillary services that are fundamentally unrelated to the normally traded product (*i.e.* the energy measured in MWh). First, a system operator must be able to control voltage to ensure system security. Voltage control requires assets that are able to generate and absorb reactive power. Second, the system operator needs access to back-start capability in order to restart the system after a black-out. Both of these functions require assets with special technical attributes that system operators do not typically possess. Such assets, owned by commercial market players, cannot be expected to be made available without specific payment. One common solution is that system operators contract these services, often through competitive bidding.

Another main category of services resembles the normally traded electricity product in MWh – and yet it is practically impossible to trade it in the same way. In the actual moment of operation, there is simply no time to conduct electricity trades to balance the system. Still, some resources must be available to respond instantaneously to changes in the system – automatic regulation. For that purpose, system operators contract for so-called "automatic reserves". These reserves respond to a technical signal – frequency – rather than an economic price signal. A very low share of total electricity supply is delivered in the form of automatic regulation. Similar reserves are also acquired to respond instantaneously to critical frequency deviations, in case of contingencies. Reserves for automatic regulation and frequency control are not products defined by the energy delivered; rather they constitute the guaranteed availability of the capacity and the technical equipment that enable adequate frequency response. Thus, automatic and frequency reserves are products principally traded in price per MW, rather than price per MWh. Again, several system operators have successfully established markets that facilitate least-cost supply of these reserves.

When an imbalance is observed, system operators will immediately try to relieve the draw on automatic or frequency reserves. This typically occurs in the minute-by-minute timeframe, in which system operators have real-time markets for system balance and bids are called on according to merit order.

Real-time markets are energy markets in which the product is priced and traded as price per MWh. In theoretic literature, the real-time market is perceived as the spot market; any markets creating commitments ahead of that – including day-ahead markets – are forward markets. Real-time balancing requires the interaction of system operators. The timeframe is still so short that commercial market players are not in a good position to ensure perfect balancing of their positions. If they were to balance individually, it would be at great loss of efficiency. The presence of system operators in the real-time segment should considerably lower transaction costs and significantly increase liquidity.

In principle, commercial market players should have incentives to make sufficient resources available for real-time balancing. However in most markets, system operators find it necessary to commit some resources as reserves, to ensure market clearing in the real-time market. Australia and the Netherlands are exceptions in that they outline only the option to commit such reserves, in case reserve margins drop below acceptable levels.

From a purely technical point of view, there may be reasons to commit reserves for real-time balancing. One is that it is difficult to acquire needed capacity at a moment's notice. Hydro plants are the power source that is truly available within the minute-by-minute timeframe required in real-time balancing. Most other plants (conventional thermal power, coal-fired and gas-fired) require some time to start up or warm up – often four to eight hours. Such plants can only be available for real-time balancing if they are partly, but not fully, dispatched. There is a real risk that a plant kept spinning for real-time balancing will not be called on even if it is available. This adds a risk premium – an opportunity cost – to the bids in the real-time market.

System operators may recognise that they can manage these risks at lower costs by contracting a minimum required amount of reserves – overall, system operators may consider that the costs for reserving capacity in advance will be lower than the average risk premiums commercial market players will charge when bidding the last resources into the real-time market. The need for reserves on these grounds will also be greatly influenced by the actual resources available in the system. Systems based heavily on hydro, such as Norway, should not need such reserves. In contrast, in systems that rely on coal-fired plants (often older ones) as the marginal resource in tight situations, a contractual up-front commitment may be a cost-effective way to ensure that an adequate part of this resource is kept spinning and available.

In the future, demand response resources and some distributed generation technologies may also add flexibility in terms of real-time market balancing. Larger generation units must decide whether or not they want to be ready for operation in real-time; the decision by a single unit may be critical for real-time balancing in a tight situation. If it does become possible to enable sufficient volumes of demand response and distributed generation resources to respond in real-time markets, it could fundamentally change the scene. A tight balancing situation normally coincides with very high demand, which implies that resources from reduced demand will probably be at their highest. Small and dispersed resources also reduce the relevance of assessments based on plant-by-plant availability (the focus in *N-1* reliability assessments). Availability becomes more of a statistical assessment, based on the expected normal response by a pool of consumers and distributed generators. As noted above, demand response in the ancillary services and operational reserve market requires both advanced metering and advanced control equipment. For most consumers, this is still not an option. However, there are examples, such as Finland, of large consumers supplying frequency reserves.

Summing up, all system operators make capacity reservations to a certain degree. System operators in several markets acquire reserves and other ancillary services through a separate market or through contracting arrangements based on competitive bidding. Such markets are in place and developing in Australia, the United Kingdom, some Nordic countries, Germany and several US markets.

Some system operators in liberalised markets have applied other strategies. Instead of contracting reserves, they have acquired the real physical assets. Svenska Kraftnät, the system operator in Sweden, acquired 640 MW of gas turbines to be used as operational reserves. This makes the independent system operator an owner of some 2% of installed capacity in Sweden. Despite this physical capacity, Svenska Kraftnät has expressed great concerns over adequate investments, particularly to meet peak demand. Its ownership of reserves does not seem to have contributed positively to investment incentives. Fingrid, the system operator in Finland, also owns its own gas turbines. It currently owns 515 MW and has decided to construct an additional 100 MW due to the increased need for reserves associated with the commissioning of the new nuclear unit at Olkiluoto. 515 MW correspond to more than 3% of installed capacity in Finland. Overall, it is likely that such ownership confuses the role of the system operator and adds uncertainty to investments, particularly in peak-load resources.

Joskow (2006) highlights several practices and features that are common in reserve markets and that effectively mute prices signals in the case of scarcity. Shortages of generation capacity result in a draw on operational reserves before other measures are used, such as load shedding. However, operational reserves often enter the market at prices far below a price that properly reflects scarcity. Joskow proposes that activation of operational reserves should be at the level of VOLL – the price cap in several energy-only markets. Another feature that effectively mutes price signals is that specific reserve requirements (such as particularly fast ramp-up times) in specific locations are often agreed outside of the market. Joskow (2006) also stresses the importance of standardisation of reserve and ancillary services products, with particular focus on allowing demand resources to participate.

Key Message

System operators must acquire and use ancillary services and other operational reserves transparently and without muting price signals of scarcity.

System operators need regulation that strictly defines their responsibility. This is also essential for defining the key responsibility of commercial market players. System operators should ensure least-cost access to required services through competitive markets dedicated to ancillary services. Ancillary services contribute to the potential revenue stream for generating and demand response resources, and therefore contribute to incentives for investments. Operational reserves used in case of scarcity should be activated without muting price signals.

Capacity Markets

It is difficult to clearly define the boundary between the strictly necessary capacity reservation for operational purposes and a market intervention to create incentives for investment. It is a grey zone. The opposite extreme is to apply measures that centrally define the total adequate level of generation capacity, and use market interventions to make sure that level is achieved. Capacity markets are one of the most extensively explored, researched and discussed issues in liberalised electricity markets. Discussions go right to the heart of how to implement textbook market designs into a real world of market imperfections and political reality, both of which can act as barriers to robust market development.

From a public policy perspective, the need for special capacity measures beyond what is strictly necessary for secure system operation should relate to the appropriate management of risks. Thus, it is fundamentally an issue of whether regulators and system operators are, in fact, best positioned to manage risks that extend beyond the operational phase. A related issue is whether total risk management costs are lower if they determine not only adequacy of resources to meet operational challenges, but also resource adequacy in general. Economic literature demonstrates that this is not the case: competing firms have incentives to make efficient investments, regulated entities generally do not. The need for capacity measures is, in effect, much more related to shortcomings in the establishment of effective markets that create cost-reflective prices and possibly to real market failures such as lack of demand response. The underlying unresolved question is this: *Are these political and possible market failures inherent in liberalised markets, in which case a capacity mechanism may be needed, or can they be mended through other means?*

The principal objective is to meet demand at all times, even when it is very high, with the least use of resources. Peak demand occurs rarely, with great variation from system to system. In energy-only markets, investors in peak-load resources must rely on rare and very high payments. This adds considerable risk that will be factored into the price, but it may also entirely deter investment. If investments are delayed to a point at which involuntary demand interruptions are inevitable, the cost to society is clearly higher than the investor's perception of the rewards to be gained by waiting. This scenario would justify a centrally co-ordinated enforcement of minimum levels of generation capacity. Price signals are critical for an energy-only market to provide adequate incentives. If regulators do not accept price spikes that reflect scarcity, despite the evidence showing that consumers can be shielded – hedged – from price spikes, an energy-only market is doomed to fail.

Adequacy of generation capacity through central co-ordination is implemented in different ways in various markets. The previous Pool in England and Wales included a capacity payment, based on a central assessment of the capacity needed to meet demand. From that, it was possible to calculate the probability to lose load in any given operational period, based on the capacity market players announced was available in the market – the loss of load probability. With a VOLL at USD 4 000/MWh (GBP 2 000/MWh), it was possible to calculate a capacity payment reflecting the dynamic value of available generation capacity. The capacity payment system was abolished in 2001 with the introduction of the New Electricity Trading Arrangements, mainly on the grounds that it was prone to market power abuse. Spain is another prominent example of the use of capacity payments and it received considerable attention in the recent white paper reviewing the performance of the market.

In the United States, most markets use capacity measures to determine adequacy of generation capacity centrally, but let the price be determined through trading arrangements. Suppliers – load serving entities (LSEs) – are given an obligation to contract for capacity, relative to the load they have committed to serve. Variations of capacity markets are implemented in PJM, New England and New York; they are under consideration in California.

Threats of market power abuse are a central aspect of the justification for capacity markets. Market clearing in periods of tight supply, including the role it has for investment at large, is discussed in previous sections of this report. The conclusion is that pricing during scarcity is critical, and that prices must clear above short-run marginal costs (SRMCs) to create incentives for investment in peak-load resources. In fact, all resources rely – at least to some extent – on revenues from prices that reflect scarcity.

In a situation in which one market player has the last available resource, this market player will indeed have full market power. Bidding to extract the full possible scarcity rent in such a situation is sometimes referred to as "hockey stick" bidding – the market player offers the last resource at maximum price (Hurlbut, Rogas & Oren, 2004). Market players with a hockey stick bidding strategy make a bet on being absolutely the last resource from time to time. In Australia, such a strategy proved risky in the long run. Price spikes attracted new peak-load resources, which eventually pushed the price downwards, even if it was still at high levels during tight situations. Considering that OCGTs are cheap and easy to build, competition should ideally never be more than 6-12 months away.

Price caps are a commonly used instrument to limit the allegedly harmful effects of hockey stick bidding. Such caps also serve as caps on revenues that should give incentives for investment – these caps on revenues have been described as the "missing money" in literature on electricity market design. Most US markets have price caps at USD 1 000/MWh, which leave very little space for market clearing in tight situations. In Australia, USD 1 000/MWh would certainly not be enough, as shown in Figure 3.4. With sufficient cross-border trade and perhaps other alternative resources, and with sufficiently low relative peakyness of demand, a low price cap may not be detrimentally distorting. This seems to be the case in Alberta and Texas, even though Texas has now decided to increase the cap. In other markets, a USD 1 000/MWh price cap also serves as a cap on the rewards necessary to provide incentives for investment. When price is not allowed to fluctuate across the full necessary range, some other mechanism must ensure resource adequacy, such as a capacity market.

In that context, a capacity obligation is also a measure to protect consumers from market power abuse. Consumers are, in effect, forced to commit to a specific type of contracting that takes on some investment risks through the capacity obligation. A central authority decides on the type of contracting that is best for consumers, whereas energy-only markets try to leave that decision to consumers themselves. The benefit of binding consumers in terms of providing certainty for investors should be weighed against alternative considerations. *First, is it possible to create a capacity obligation without losing some of the efficiency gains that a more decentralised investment decision process brings?* Investment incentives and long-term efficiency are discussed in previous sections, where it is shown that maintaining capacity adequacy through a centralised planning process puts the benefits from introducing competition in jeopardy, particularly if reliability criteria are not adapted to the new competitive framework. *Second, does a capacity measure remove market power abuse or simply move it from one market segment to another? Finally, are energy-only markets fraught with market power abuse?*

In reality, capacity measures have so far tended to move market abuse rather than remove it. The capacity payment system in England and Wales was undermined by market power abuse. Research literature provides considerable evidence of market power abuse by withholding capacity to increase the capacity payment (Evans and Green, 2005). PJM conclude in their state of the market report for 2005 that capacity market results were competitive but market power remains a serious concern, due to high concentration (PJM, 2006).

Market power is also a great concern in several energy-only markets. The Danish Competition Authority ruled in 2005 that Elsam, an incumbent generator, had abused its dominating position in 2003. The Authority investigated the case on its own initiative, after the system operator expressed concerns (The Danish Competition Authority, 2005). Nordic competition authorities issued a report in 2003 expressing general concerns about the level of market concentration (Nordic Competition Authorities, 2003) and specifically the effect this concentration may have on overall price formation. In reality there are few clear cases in which market players have abused their position during situations of real scarcity to bid into the market at extreme prices. In fact, there are very few examples in the Nordic market of prices above USD 1 000/MWh. Australia is also concerned about market power in situations of scarcity, but average prices are very low. Scarcity pricing may have resulted from market power in specific half-hour trading sessions but, in effect, it attracted new competition, mainly through investment in gas turbines.

Price caps are unlikely to be an effective instrument to mitigate market power abuse and the consequence of distorting investment incentives may be detrimental to long-term efficiency. However, price caps probably do serve the purpose of creating acceptability amongst consumers and their political representatives who may not accept extreme price spikes. Again, Australia demonstrates that price spikes do not imply high prices. Most consumers pay bills that are averaged over several days, months and perhaps even years. In that sense, a capacity obligation is merely an obligatory form of contracting enforced upon consumers on top of the array of contracting arrangements at their disposal. It is doubtful whether such obligations, as opposed to freedom of choice, effectively protect consumers and optimise consumer welfare.

It is sometimes argued that voluntary contracting to manage risks is not sufficient to ensure adequate investment in power generation capacity, particularly in light of the long life times of generating units, the often high capital costs and the long lead times.[15] In other industries, such as pharmaceuticals, computer chips and shipping, very large investments are made with no or very little contract backing, and often with uncertainty about basic demand. Considerable certainty about basic electricity demand should, in fact, substantially lower the risks of investing in power generation as compared to many other sectors. Again, the key issue seems to be a failure to create a market design that provides incentives and rewards to those willing to take the inherent risks, and that enables effective risk management through contracts.

At present, lack of active demand participation is the critical market imperfection that drives debate on the need for capacity measures. Lack of demand response to balance scarce resources may be a root cause of market power abuse and alleged risk management deficiencies. A two-sided market, with active participation both from supply and demand, is more robust and more likely to lead to competitive outcomes, even when supply is tight. A demand side that is inherently price inelastic may be the critical market failure that makes capacity measures inevitable. In reality, demand is clearly elastic; its pattern can change depending on rate structure. This is true intuitively, and there is now early evidence that transaction costs are manageable for some consumers and for some types of consumption. In terms of assessing the need for capacity markets, the critical issues are then: *What are the necessary demand response volumes and characteristics to make the market robust? How can incentives be created to encourage demand response in capacity markets?*

Price is determined by many factors; demand is only one. Availability of plants and transmission assets may also have great effect in individual hours. Nevertheless, an examination of the situation in New South Wales shows that very little demand response, as a share of total demand can have a

---

15. *Bushnell (2005) provides a good overview of arguments for using capacity markets, including the need for contracting to back investments with long economic lifetimes.*

significant impact on system balance. In New South Wales in 2005, average prices in the 10 hours with the highest prices were some USD 5 000 MWh (AUD 6 707/MWh). Average prices in the 10 hours with the next highest prices were some USD 2 000/MWh (AUD 2 407/MWh) – i.e. were lower by more than USD 3 000/MWh. Average demand in the corresponding hours differed by only 3 MWh. In the next 10 hours of the price duration curve, the average price drops to some USD 500/MWh (AUD 583/MWh) and average demand falls by 158 MWh compared to the previous 10 hours. The orders of magnitude of prices and corresponding demand in this case study illustrate the potential of demand response. Demand response resources corresponding to just 1% of peak demand may considerably lower prices and may offer critical competition for the last remaining generation resources.

Spiking spot prices in the NEM, as illustrated in Figure 3.4, also indicate that the Australians may have found the "missing money" – the money that is missing in markets where prices and revenues are capped. Assuming that marginal costs of an OCGT are AUD 100/MWh[16], in 2005 a 100 MW unit would have earned about AUD 20 million in New South Wales and about AUD 7 million in South Australia (after deducting fuel costs). There is, of course, great risk. The revenues less fuel costs varied between about AUD 3 million and AUD 26 million over the 2000-05 period in South Australia. Total revenue less fuel costs in the same period would have been AUD 80 million. Assuming that such a unit would cost about AUD 50 million in overnight investment costs, it would have been paid back in five years. The NEM seems to provide ample incentives for investments in OCGTs, even though this market's average prices are very low.

Demand response and its potential are discussed in previous sections. It is emerging, but only slowly and is still far below its likely potential. This calls for government and regulatory action, but are capacity measures the right kind of action? It is possible to commit consumers to respond in ways that make them eligible for a capacity credit; the credit can, in turn, fulfil capacity obligations. In the PJM, load serving entities (LSEs) can reduce their obligation by committing consumers to active load management (ALM). In turn, ALM resources must fulfil a set of requirements to be eligible including load reduction during six consecutive hours, with two hours notice and throughout a PJM planning period (one year). Consumers with some load that fulfils these conditions are able to extract some of the value of the corresponding reduction in capacity obligation. In 2004, some 900 MW were committed as ALM resources.

---

16. *This is based on assumptions on gas prices used in Table 2.1. Wholesale gas prices in Australia are not so transparent but they were considerably lower than the USD 6/MBtu assumed here. Fuel costs of USD 40/MWh may in fact be more reflective of the real costs in 2000-05.*

Consumers with potential demand response resources that are not committable under such conditions must find other ways to extract value. PJM runs two other programmes: the Emergency Load Response and the Economic Load Response programmes. Various categories of demand response are eligible within these programmes, including load reduction that is committable only one day ahead. There are no long-term commitments and consequently no capacity value. All payments are relative to the actual energy delivered and the actual locational marginal prices (LMPs), with the exception that the Emergency Load Response programme offers a minimum price of USD 500/MWh. Demand response at costs higher than the price cap of USD 1 000/MWh are thereby cut off. In 2005, maximum LMP was USD 287/MWh; the price exceeded USD 200/MWh during 35 hours. Demand response is likely to cost substantially more than USD 200/MWh in many cases. Peak prices at such low levels probably also accurately reflect the fact that the capacity obligation ensures that the system is never particularly tight. Consequently, there is no need for demand response resources.

Capacity payments through a capacity market or through ancillary service markets can contribute considerably to the development of specific types of demand response resources. But if the full array of demand response is to have incentives, effective pricing of scarcity is inevitable. Low price caps and other out-of-market practices that effectively mute price signals are serious barriers to that end.

The concept of capacity markets faces another fundamental challenge in terms of meeting its objectives without undue loss of efficiency. Demand is not linked to specific resources of supply in open and competitive markets. Generation in one area may meet local demand, but may also meet demand in a neighbouring interconnected market. Neighbouring markets will benefit from local capacity obligations without carrying the obligation to share the costs. Such "free-riding" jeopardises the effectiveness of local capacity markets in ensuring adequacy. Eventually the obligation may have to increase to reflect export-capacity. Local capacity markets may also effectively fend off competition from neighbouring markets.

Capacity markets attract investments relative to neighbouring energy-only markets; long-term efficiency and reliability would be undermined if they were to live together. The EU Security of Supply Directive introduced the option to implement a capacity measure based on open tendering. It is difficult to see how individual interconnected systems considering such measures can avoid free-riding and inefficiency in the longer term.

Existing capacity markets have largely served their main objective of ensuring adequacy of generation capacity. In fact, the most prominent and long-living capacity markets are in the North-eastern United States, a region that experienced an investment boom in 2000-03. It must not be overlooked that this investment response left bankruptcies and defaults in its wake.

Capacity markets have developed continuously to improve and adapt to changing circumstances, as is also the case for all other liberalising electricity markets. However, capacity markets have demonstrated some more critical deficiencies. The most serious short-coming is that they have not eased market power problems in concentrated markets – in fact, they have perhaps exacerbated the problems. These problems originate from the same source as in energy-only markets: namely, demand is more or less fixed and predictable, as is installed capacity (at least for 6-12 months). Thus, when the capacity credit is earned through installed capacity, both supply and demand become more or less inelastic. Even a few MW can significantly change the price; thus, generators with market power have a strong incentive to withhold capacity. The PJM initially based the capacity obligation on installed capacity – the installed capacity market. The obligation is now based on an assessment of actual availability – the unforced capacity credit market. Capacity credits are awarded to capacity to the extent it is available during peak-load summer periods. Credits are adjusted down with an average availability factor, computed on a rolling 12-month basis. Unforced capacity credits have an advantage over installed capacity in terms of creating incentives for efficiency: The fact that the total supply of unforced capacity credits changes dynamically every month probably also reduces – at least somewhat – the scope for market power abuse.

System operators and researchers in the United States continue to search for solutions to improve performance of capacity markets. Proposed models include locational elements, they enable competition from new plants before being built (thereby also offering a hedging opportunity for investors), they introduce a sloping demand curve, and they allow for longer term contracting. Box 3.4 describes one of the latest advancements in research, which includes many of the elements from earlier models and addresses many of the difficult challenges of capacity markets and energy-only markets. One of the most notable features in this model, proposed by Peter Cramton and Steven Stoft, is that it includes an element of real scarcity payment (Cramton and Stoft, 2006). With that feature, the model may meet its main objective of resource adequacy without simultaneously creating a barrier for the long-term development of a robust energy-only market.

Box 3.4 . Transition to robust energy-only markets: a proposed capacity measure that may not be a barrier

Peter Cramton and Steven Stoft published a paper on capacity adequacy for the Californian Electricity Oversight Board in 2006 (Cramton and Stoft, 2006). The title, Convergence of Market Designs for Adequate Generation Capacity *indicates that it comprises several of the preferable features proposed in recent research on capacity markets. The new model builds on the assumption that supply and demand are given, and that demand is inelastic. In such a framework, unacceptable market power is inevitable in the few hours in which the last resource is required. In other words, it is impossible to avoid the unnecessarily high risk premium imposed on consumers for covering the risk of investments in peaking units. Thus, imposing price caps is a necessary response to protect consumers. This in turn caps the revenues needed to provide incentives for peak investments - the "missing money". Eventually, this domino effect may put system security in jeopardy.*

The feature that sets apart this specific model and takes the concept of capacity markets forward is that it does not mute, but rather magnifies price signals. The price magnification feature is constructed by a system operator, and will hence only be as accurate as the system operator is able to make it. In short, it reverts back to a fully centralised process in terms of determining the total adequate level of capacity. This model assumes that full knowledge about supply and demand can be acquired, and thus there is less risk that it will result in a loss of efficiency: a central planner's dream. It will deliver substantially different investment outcomes than a regulated industry only if reliability criteria are comprehensively updated to the new competitive framework. Under such conditions, the model provides a good overview of many of the advancements in capacity adequacy research and may point out a direction for future developments.

The first step in the model is to determine the total adequate level of installed capacity, a task that can be performed by the system operator. An obligation is then put on all load serving entities to purchase a share of the total capacity credits that corresponds to their share of load. This is like the installed capacity market (ICAP), with one important difference: the credit is not for capacity in the coming year, but for three years ahead in time, thereby offering new investors a hedging opportunity. The money received in the capacity auction does not flow directly to generators. A performance incentive is introduced, which may affect the payments. The incentive element is in the form of constructed scarcity revenue, and the scarcity rent is defined by market prices that are above a certain strike price.

The strike price is set at the variable cost of a benchmark peaking unit – say USD 100/MWh. Scarcity revenues are calculated when market clearing prices are above the strike price, and as the share of the price above the strike price, times a spot price magnifier. The magnifier is calculated so that prices are magnified to levels that ensure the total scarcity revenues are sufficient to finance peaking units. The model thereby corresponds to a competitive energy-only market. Each generator's share of the scarcity revenue is calculated based on installed capacity. If a generator actually supplies capacity that corresponds to its share during the hours of scarcity, it will receive only its share of the capacity auction revenue. Any deviations in actual performance will be added to or deducted from the share of the auction revenue. Thus, a generator supplying more than its share of total installed capacity will receive a bonus, at the expense of generators supplying less. Additions match deductions, which means that the only payments for consumers are the energy payment at prices up to the strike price, and the capacity payment. In that sense, the model resembles a call option, similar to a model proposed by Shmuel Oren (Oren, 2005). The call option leaves the consumer fully hedged to price spikes, but not to variations below the strike price.

Most capacity models tend to act as barriers to transition. The Cramton and Stoft model has various built-in features to avert this short-coming. The most critical feature is that it uses a constructed scarcity price signal. In principle, demand response can be contracted to benefit from scarcity pricing. This would probably require that capacity obligations be calculated without accounting for demand response, thereby forcing demand response into firm commitments to become eligible. It is still a challenge to nurture the full array of potential value-adding demand response categories. With fully hedged prices, it is proposed that price caps can increase without harming consumers. With increasing demand response, there will be more prices above the strike price. This will reduce the need for the magnifier, eventually letting it fade away.

Cross-border trade is not reflected in the model. In fact, it is regarded as a threat to system security during periods of shortage. For the case of California, it is suggested in the model to be forbidden. Sharing of resources across borders, particularly when they are scarce, is an important efficiency benefit from efficient trade and well co-ordinated system operation. The model effectively rules out this opportunity in its basic design, which is a serious shortcoming.

Capacity Safeguards to Ensure a Safe Transition

The need for capacity measures to give incentives for investment is often discussed with reference to a static snap-shot of a market, with all its current shortcomings and challenges. Many of the elements that contribute to the need for capacity measures are driven more by political factors than by deeply rooted market imperfections. Political factors change. Thus, it is important to consider the dynamic effects of a capacity measure, and the effects it has on the development of the market. Some of the alleged market imperfections are likely to be of only a transitional nature. It is therefore critical that a capacity measure does not stand in the way of a healthy transition to a robust market. Still, some sort of capacity measure may be necessary to get through a transition, at least from a political point of view.

Many liberalising markets are likely to face a transitional problem in the development towards more robust markets. Liberalisation is a process that takes time: effective trading arrangements, liquid contract markets and genuine retail competition do not emerge over night, but rather evolve over several years. Demand response is, in several markets, the critical resource required for the development of robust markets, particularly when cross-border capacity is limited and relative peaky-ness is high. Demand response also relies on effective trading arrangements and competition. Moreover, demand response is likely to develop only when the resource is actually needed – *i.e.* when the system has shown signs of tightness. Demand response emerges with volatile prices that spike to high levels from time to time; regulatory enforcement of protection of consumers from price volatility is also an effective barrier for demand response.

It is understandable that governments and regulators want to protect consumers from market power abuse when the market is still fragile from lack of demand response. More critically, demand response may be the resource that ensures reliability, particularly in systems with extreme but rare load peaks. Rolling blackouts and jeopardised system security due to fragile market conditions are not acceptable market outcomes.

Most markets have implemented some sort of regulatory intervention to overcome the transition towards more robust markets. For the most part, softer regulatory instruments have been used as compared with capacity markets.

Most markets reserve capacity in various forms to serve as a safety net. Australia is an example of perhaps the softest form of intervention. NEMMCO, the Australian system operator, can tender for reserve capacity through a reliability safety net procedure. A reliability panel assesses the minimum required level of reserves, and NEMMCO can tender for reserves when available capacity drops below these regional minimum levels. In 2004, the reserves of available capacity for expected peak-load corresponded to only 3% of the peak-load. NEMMCO used the safety net for the first time in 2006, procuring 375 MW of reserves in Victoria and South Australia. Some 125 MW was from a company, Energy Response Pty Ltd., which aggregates demand response.

Reserves bought in the Australian safety net procedure effectively resemble operational reserves bought in most other markets (*e.g.* spinning reserves in the PJM, minute reserves in Germany, disturbance reserves in Sweden and standing reserves in Britain). The Netherlands is an exception. TenneT, the Dutch system operator, relies on the real-time energy market to acquire balancing services, without longer term reserve commitments. TenneT also have a safety net possibility to acquire additional reserves, similar with NEMMCO's, but this option has not been found necessary.

Norway is another interesting example of relatively softer market intervention. Statnett, the Norwegian system operator, has established a market for operational reserves – the regulating power option market (RPOM). This is a weekly market in which capacity is reserved to bid into the regulating market; up to 2 000 MW is acquired. The market operates only during the winter season, when demand is high. It is thus not a reservation of capacity that is necessary due to technical constraints. Norway relies 100% on hydro power, which does not need any warm up time. Even if the RPOM may distort the energy market, it has had at least one important effect in supporting the transition to a more robust market. RPOM has attracted substantial amounts of demand response resources. During the winter 2005/06 at least half, and often more than two-thirds, of the total volume acquired was from the demand side. The RPOM has helped develop demand response resources, also to the benefit of the market at large. Significant demand response played a critical role in balancing scarce energy resources through the Nordic drought in 2002/03.

Droughts are still a great concern in Norway, particularly considering that new investments are highly controversial. Gas-fired generation is challenged for its greenhouse gas footprint compared to hydro; new hydro and wind developments are challenged for their impacts on landscape. Political and regulatory uncertainty have delayed necessary investments. One response from Statnett was to build a new 700 MW transmission cable connecting Norway with Netherlands, which is due for commissioning in the winter 2007/08. A new cable can only postpone the time when new investments will be needed; it does not solve more local shortages. Future droughts are thus still a major concern, even if the Nordic market managed the 2002/03 drought efficiently (IEA, 2005a). The Norwegian regulator has now authorised Statnett to operate an energy option trial programme, along similar lines as the RPOM. It is intended as a safety net in case of extreme drought. It will run as a trial for the winter 2006/07, to be evaluated after one year. Only demand is eligible to bid into this energy option market.

New Zealand is another hydro-dependant country with energy concerns during droughts. It does not rely 100% on hydro, but it is electrically 100% isolated. New Zealand operates an energy-only market with considerable success. But two severe droughts in 2001 and 2003 put pressure on the system. Emergency conservation campaigns, which cut 10% of demand, were implemented to manage the situation. The 155 MW oil-fired unit, Whirinaki, was commissioned in 2004 by the Crown, to be used in case of shortages (IEA, 2006).

Sweden uses a model for acquiring reserves which is another step into the grey zone between energy-only markets and capacity markets. This model is intended to be transitional, operating from 2003-08. Mothballing and possible closure of two larger oil-fired plants in the late 1990s forced Svenska Kraftnät, the Swedish system operator, to contract for capacity reserves in 2000-02. This triggered an in-depth inquiry by Svenska Kraftnät into generation adequacy in liberalised markets – the first comprehensive study of that kind in the Nordic market (Svenska Kraftnät, 2002). Svenska Kraftnät took the decision to acquire up to 2 000 MW in open tender in a deliberate move to implement a simple model that would smooth, but not hinder, the transition to a more robust market. Significant emphasis was put on demand response resources. Svenska Kraftnät managed to buy increasing shares from the demand side; in the tender for the winter 2006/07, some 564 MW of the

1 989 MW acquired were from the demand side. Svenska Kraftnät maintains overall control and decides when the reserve should be activated. It is then bid into the market at USD 1 000/MWh (SEK 8 000/MWh). The maximum accepted bid in the Swedish market is SEK 50 000/MWh. Svenska Kraftnät is currently considering whether the transitional model has actually created transition. In a report to the government, the operator now assesses that capacity will still be tight in extreme weather conditions after 2008, but will be somewhat relieved in a little longer timeframe if current investment plans materialise. Svenska Kraftnät prefers to end the model, but supports a plan from the Association of Electric Utilities in Sweden to prolong the system for some years under the responsibility of commercial market players (Svenska Kraftnät, 2006). Finland is now working on implementing a system along similar lines, also triggered by intentions to retire oil-fired power plants.

Key Message

Capacity measures may be necessary in a transition to robust markets, but they are a barrier if used to mute market signals, for example, through price caps.

Robust markets rely on effective regulation and trading arrangements. Dynamic market interaction, cross-border trade and demand response take time to mature. System reliability may be fragile during the process and transitional capacity measures, acting as a safety net and without muting appropriate market signals, may be necessary. Consumers are best protected in the long term with competition, not with low price caps that mute market signals and necessitate permanent capacity measures.

ENABLING INVESTMENT THROUGH POLICY CLARITY AND REGULATORY EFFICIENCY

IEA member countries are increasingly relying on competitive markets to ensure security of energy supply at the lowest cost possible. The market reform process is expected to continue with the goal of establishing a clear definition of – and indeed division between – the roles of governments, regulatory authorities and stakeholders. As discussed in previous chapters, inefficient regulation, which often implies regulatory delays in power plant licensing, can add significant cost to a project, and can be a serious barrier to investments. In addition, the lack of a stable, clear and predictable policy and regulatory framework is a major source of uncertainty and risks to investors, who need to take a long-term view in project planning and assessment. Furthermore, concern over lack of competition is another key issue that governments and regulators need to address. Thus, governments and regulators play a key role in enabling investments in power generation. This role should include the following elements: ensure that the policy framework supports the efficient development of robust, competitive wholesale electricity markets; provide to markets a clear, stable, and predictable policy and regulatory frameworks; and improve regulatory efficiency. The latter implies that governments should also ensure that regulators can operate independently from the interests of all stakeholders (including government) and that regulatory authorities have adequate resources and competencies to fulfil their mandate.

Building on the preceding chapters of this report and other IEA work, this chapter identifies the main features of effective regulatory frameworks; discusses various approaches in addressing public concerns related to power plant licensing; and assesses models of nuclear plant licensing in various IEA jurisdictions.

It should be reiterated that transmission is the conduit that delivers generated power to markets. Thus, it is imperative that potential reform of power plant licensing be undertaken in conjunction with reform in transmission. In general, siting of transmission is more challenging than siting of generation, primarily because transmission often spreads across several jurisdictions. Poorly planned transmission can create serious barriers to investments, particularly if investors foresee problems related to grid interconnection and transmission congestion.

Each IEA member country develops policy and regulatory approaches in the context of specific and unique national circumstances. Approaches that are effective in one country may not be applicable in another. Thus the following survey of national regulatory approaches should be considered as "good" practices, not necessarily "best" practices.

Establishing an Effective Regulatory Framework ....

In the evolving market environment, electric utilities and other market participants are constantly challenged by market restructuring and increased competition, as well as by high and volatile fuel costs and tightening environmental regulation. At the same time, security of energy supply has risen to the top of the policy agenda. In this evolving market environment, policy makers and regulators play a key strategic role in ensuring that the policy and regulatory frameworks facilitate rather than deter timely and sufficient investments. This does not necessarily imply a more interventionist role for government; it rather suggests that government should focus on reducing uncertainty and improve the timeliness, predictability and consistency of decision making. Policy makers, more than ever, need to review and adapt the institutional and regulatory frameworks to reflect new market realities. They must also strive to find the best strategies to enhance energy efficiency and address environmental concerns, while ensuring long-term generation adequacy at the lowest cost possible.

Effective regulation requires regulators to be independent – in two dimensions: politically independent and independent from stakeholder interest. Political independence means that regulators are shielded from short-term political influence. This can be achieved through irrevocable mandates for regulators and through other measures such as separate budgets and autonomy in human resource management. Political independence is particularly critical in the case of state-owned electric utilities in order to avoid conflicts of interest between the state as owner and regulator. Independence from stakeholders implies that regulated parties have limited influence on regulatory decisions. It serves as a mechanism to ensure that regulation is fair and objective, and does not favour one group of stakeholders over the others. Independence from government is particularly important in that it builds confidence amongst investors in the stakeholder community.

Market transparency is another key element of effective regulation. Transparency can be actively pursued, for example, by providing the information market participants need to make informed investment decisions. Transparency is prerequisite for good risk management – an area in which public policy plays a critical role. Well-designed liberalised markets provide a certain degree of transparency, but the level needed to support risk management can only be achieved through timely, relevant and easily accessible information to all market players. Lack of transparency makes it difficult to manage risk, and such a limitation will likely distort incentives for efficiency and hinder the development of competition. Transparency facilitates electricity trade and enables market players to operate independently of old incumbents.

Transparency is deemed essential in the following critical areas: *i)* Market rules, including those associated with electricity pricing and transmission rate setting; *ii)* Current state of market, focusing on market fundamentals; and *iii)* Longer term market analysis. The challenge is that, in most cases, the necessary flow of relevant information does not happen voluntarily. Regulatory intervention may be required to ensure that information is disseminated to the marketplace and is also easily available and accessible.

The overall aim of market rules is to create a level playing field; *i.e.* to give all market players real and equal opportunities to actively participate and to interact as efficiently as possible. Effective market rules and design do not develop by chance; they are the result of careful consideration of costs (transaction costs) and benefits. This requires thorough examination of constraints such as demand fluctuations, start-up time, security limits, intermittency, coordination between system and market operators for cross-border trade, etc. Regulators should make these rules as clear as possible, and ensure they are widely understood by all market participants, and are impartially and predictably enforced.

Information about demand and the availability of transmission generation capacity is critical to enabling analysis and understanding of the state of the electricity system, both in the operational phase and for the longer term. If market participants do not have access to the information needed to analyse and understand the state of the system, they cannot respond appropriately to market needs. Transparency and easy access to fundamental information are still under development in several markets. Lack of, or slow progress towards, a minimum level of transparency is a key barrier to appropriate investment response in competitive markets.

Several jurisdictions now require publication of information to provide transparency. These include the United Kingdom, the PJM (North-eastern United States), Alberta (Canada) and Australia (IEA, 2005a). Transparency is at the top of the agenda in the current European market development. The European Regulators Group for Electricity and Gas (ERGEG) has prepared guidelines for transparency and information management in electricity markets (ERGEG, 2006), which seek to establish a consistent approach across member states (Box 4.1). This effort has already prompted industry to pursue certain initiatives to improve transparency. For example, in late 2006 the association of European Transmission System Operators (ETSO) launched an information platform that includes data on the availability and use of transmission systems across all member systems.

Box 4.1. European regulators' guidelines on transparency

The European Regulators Group for Electricity and Gas (ERGEG) was established by the European Commission (EC) to facilitate co-operation with and seek advice from regulators across the European Union. To ensure the development of an efficient marketplace, ERGEG is creating a series of guidelines that reflect its opinion on good practice in implementing the EU Electricity and Gas Market Directives and regulations. Since 2004, ERGEG has focused part of its work on guidelines for information management and transparency in the electricity market. In the process of developing the guidelines, ERGEG sought input through public hearings and discussions. One of the platforms for public consultation is the annual Florence forum hosted by the European Commission to generate momentum in the development of an internal electricity market.

The guidelines include the following minimum information requirements to ensure a transparent market (ERGEG, 2006):

- **System load by control area**
  - *Both ex-ante (forecasts) and ex-post (actual) values by control area and over various timeframes (hours, days, months and years). Also including forecast margins relative to installed capacity.*

- **Transmission and access to interconnections**
  - *Network investment and planning, covering expansion proposals, planned works and outages.*

  - *Capacity allocation and management, including forecasts of interconnection capacity (week and month ahead), capacity requested, actual utilisation and congestion income. Also including the scheme for calculating total transfer capacity and the transmission reliability margin.*

  - *Network operation, including ex-post information on actual outages, realised physical flows and average hourly physical flows versus thermal ratings.*

- *Generation*

  - *Installed and available capacity, both current and future plans up to ten years ahead, at minimum on an aggregate base by primary fuel source.*

  - *Scheduled unavailabilities ex-ante, including start and stop dates of outages and capacity involved.*

  - *Scheduled generation ex-ante, on an aggregate base by control area.*

  - *Hydro reservoir filling rates per week and relevant geographic area.*

  - *Intermittent generation such as wind power, both forecast and actual generation.*

  - *Actual availability of capacity ex-post.*

  - *Actual generation ex-post, by control area and primary fuel source.*

The development of these guidelines is crucial to creating transparency. At present, various issues related to their enforcement remain unresolved. This, of course, has important implications for the successful implementation of the guidelines.

Information about the current state of a power system feeds into analysis and understanding of long-term market development. Market participants can use this information to analyse needs for new generation capacity, as well as to assess risks and rewards for specific investments. Most countries have recognised the importance of this information. Good examples can be found in the Statement of Opportunities published by NEMMCO (Australia) and the Seven-Year Statement by the National Grid (United Kingdom). These annual documents describe the current state of the power system and the prospects for the future, focusing on both demand (including peak demand) and supply (including tracking of new generation projects). ETSO compiles an annual assessment drawing on regional input from across Europe. In addition, the EU Directive on Security of Supply gives member states a mandate to monitor market developments. The North American Electric Reliability Council (NERC), which has evolved into an Electric Reliability Organization (ERO) with mandatory standards, compiles annual assessments submitted by the regional reliability councils with representation from American and Canadian entities. The *US Energy Act of 2005* includes specific requirements for the ERO to analyse and publicise the state and prospects of the power systems.

The need for co-ordination between transmission and generation investments provides strong incentive for mandating system statements. Because transmission investments remain largely regulated, it is essential that these investment decisions be as transparent as possible. This is particularly true for investments aimed at removing major transmission bottlenecks, which can fundamentally alter the profitability of a specific generation project. Mandated system statements can also help to lower entry costs for smaller generators that are newcomers to the market, thereby potentially improving the level of competition.

In Europe, system operators are usually responsible for compiling system statements – a logical choice given their intimate knowledge of the system. However, system operators may not necessarily have the right incentives to ensure that the analysis is truly objective. In many cases, system operators also own transmission. Thus, it may be in their interest to present a biased analysis in favour of, for example, further transmission development or an affiliated incumbent generator and/or retailer.

Basic statistical information about electricity consumption and generation is traditionally collected and publicised by governments, government agencies and industrial bodies. The IEA collects such statistics from official sources representing member governments as a means of enabling regional and global compilation and analysis. In several countries, this process has eroded following the liberalisation of electricity markets. Individual market players now have different commercial incentives and often want to protect information that was previously considered uncontroversial. Knowing that liberalisation changes incentives, it is appropriate to re-examine roles and responsibilities. Governments have a critical role to play in adjusting responsibilities to ensure that high quality basic statistical information continues to be collected in a timely manner.

A key role of regulators is to ensure the responsible development of an effective energy infrastructure, a challenging task that involves balancing diverse interests. Regulators should support market-based instruments to facilitate private investments in competitive markets. At the same time, it would be appropriate for regulators to demonstrate diligence in regard to the potential socio-economic and environmental impacts that markets may not fully address, *i.e.* externalities. In exercising their mandate, regulators should ensure that public concerns are identified and understood in the early stages of project development, and that the correct balance is achieved among economic, social and environmental factors.

Promoters of energy infrastructure projects (including power plant facilities) often face long lead time to obtain regulatory approvals. In many cases, this is attributed to the existence of multiple layers of regulatory bodies, operating at both central and local levels. While each entity pursues its own mission, it may be the case that jurisdictional mandates are poorly defined, which creates potential for duplication and for administrative gaps that result in inadequate attention being devoted to some aspects of investment project review. Such fragmentation is a key challenge for today's regulators: it makes it much more difficult to achieve integrated decisions on projects that reflect the overall public interest. Fragmentation is also characteristic of interconnected regions, such as those in Europe and in some United States–Canada markets. Regulation should aim to lessen the impact of regulatory fragmentation through harmonisation. The United States has taken steps to improve harmonisation by establishing the Nuclear Regulatory Commission and giving it overall responsibility for licensing new nuclear power plants. In some cases, harmonisation of rules will not be sufficient, and regulatory efficiency could be better achieved through inter-regional regulatory structures, such as those established in Australia and in the United States.

Cross-border issues can present significant challenges for both regulators and investors. Because regulation can vary from one country to another, investors trying to comply with multiple national regulatory frameworks often need to duplicate their efforts. This can lead to additional delays and increased costs. On the regulatory side, the process of assessing whether a project is in the public interest may be further complicated by public opposition originating from external sources. For example, plans to install a nuclear or coal-fired power plant near national borders may lead to actions or protests by a community that is actually situated in a neighbouring country. Effective regulation requires harmonisation of regulatory processes to the greatest extent possible. For projects with cross-border implications, coordinated and harmonised regulatory actions are highly desirable.

In addition to ensuring that markets have timely and adequate capacity, an efficient regulatory framework must also properly address issues of overlap and duplication, particularly relating to processes and timelines for multi-jurisdiction projects. This can be pursued in a number of ways including setting legislated time limits, establishing a major projects office, and identifying and empowering a lead agency to set timelines. In addition, it could be useful to set out timelines in the terms of reference of joint review panels and to establish institutional links between different regulatory authorities.

Recent years have seen significant investments in renewable-based generation. Many renewable projects, especially wind power, are in remote locations far from load centres. In many markets, access and interconnection to transmission and distribution systems can act as barriers to investments in renewable and distributed generation projects. In liberalised markets, access to transmission systems must be available under fair and non-discriminatory terms and conditions to all market participants. These barriers to investments should be alleviated and removed. Thus, regulators can play a key role in facilitating open and non-discriminatory grid access while ensuring that access and interconnection rules are clearly established and stable.

In reality, network access often remains limited or problematic as a result of inadequate market rules and standards, or due to connection procedures that are complex and, sometimes, costly, time consuming or unfair. Because of their small size and/or location, many renewable and distributed generators need to be connected to local distribution networks rather than the national transmission network. Distributed generators (DG) in some markets have expressed concern that projects are unduly delayed because it is not possible to establish quick and easy connections. Regulators have a role in providing incentives to utilities to support and facilitate DG connection.

More importantly, regulatory frameworks should be adjusted to truly create a level playing field – i.e. one that does not discriminate against non-utility generators. It seems steps must be taken to eliminate possible resistance from utilities. This may require the establishment of a regulatory regime that offers incentives to promote and reward the development of improved efficiency systems. This may involve substantial changes in the way distribution networks are designed, organised and financed.

Regulatory efficiency stems from or can be associated with "smart regulation", a concept aimed at improving regulatory systems in order to keep pace with evolving market realities and societal needs. It strives for a better co-ordinated, more transparent system that is forward thinking and accountable to citizens. Smart regulation aims to strengthen the policies, processes and tools needed to sustain high levels of regulatory performance and facilitate continuous improvement. It is a key driver behind efforts to harmonise and/or adopt coordinated – and even uniform – regional approaches to regulation. Australia provides a good example of the kind of initiatives needed to enhance regulatory efficiency: It has rationalised its regulatory system by replacing 13 regional regulators with one national regulator for electricity and gas.

Regulatory efficiency also involves developing an integrated approach to energy regulation and competition regulation. As discussed previously, the lack of competition in a given market can be a serious barrier to investment in new generation. One way to achieve integration is to develop close working relationships between energy regulators and competition authorities, so that these entities are fully aware of each other's policies and objectives.

Key Message

Investments in power generation are long-term. Therefore, markets need clear, stable and predictable regulatory processes and policies. Governments and regulators play a key role in enabling investments through clear and stable policy and efficient regulation. Regulatory efficiency requires transparency and independence. It also implies more uniform regulation and closer links between competition and energy regulators.

Regulators should strive to balance their dual role of enabling investments and protecting public interest. Markets rules should be clear and consistent, and provide for a level playing field. Market-based prices, e.g. locational marginal pricing, should provide adequate signals for investments and attract investments where and when they are most needed. Uncertainty is a market reality that implies risks for investment in power generation. Regulators must establish an effective market framework in which risky investments can occur and can deliver appropriate reward. Regulators should contribute to transparency enhancement in all segments of the electricity supply chain. Governments need to improve the regulatory processes through clearly defined and shorter time lines for project approvals, and by implementing a faster, more cost efficient and more predictable process, particularly with respect to facility and site approvals, and other procedures.

Addressing Public Concerns in Power Plant Licensing ....

The construction and operation of power plants can have significant impacts on individuals, groups, local and/or regional communities; this is true for many types of generation projects. The impacts often depend on the size of the project, and the type of technology to be applied. For example, the installation of wind farms may reduce the amount of land available for recreational or agricultural purposes. The operation of a coal-fired power plant will contribute to higher emissions. Thus, there is growing public interest in safety, environment and land use issues especially in the context of regulatory approvals (or licensing) of generating facilities. In this context, investments in power generation projects and related facilities have attracted increasing attention and, in many cases, opposition from concerned citizens and other interest groups. In some cases, the projects have been delayed, scaled back, or cancelled altogether.

While public inputs are a critical component of regulatory process, there is a tendency amongst the general public to react to new generation investment with a "not-in-my-backyard" (NIMBY) attitude. In some cases, the response is more extreme, with a "build-absolutely-nothing-anywhere-near-anybody" (BANANA) or even "not-on-planet-earth" (NOPE) attitudes. These responses reflect increased environmental concern and increased value of private property.

In this report, the phrase NIMBY is used in a narrow sense to express a form of public opposition to the construction and operation of generating facilities, including nuclear power plants, in their neighbourhood. This movement exists around the world, and is based on reasons ranging from health and ecology to aesthetics and property value. NIMBY groups are not necessarily against the project; they simply do not want the project to be implemented in their immediate neighbourhood. The local aspect of these movements needs to be clearly understood in developing appropriate regulatory approaches for plant site permitting.

Some groups are highly effective with strong resources, a high level of technical expertise and an ability to tackle the complex issues of power plant construction and operations. In effect, they have developed a strong capability to assess the relevance of projects, based on their own perspectives and interests. Local public concerns can make the licensing process of new energy infrastructures more time-consuming and costly. They can also result in increased delays, higher costs and greater uncertainties for project development. Long inquiries can be expensive for both the investor and for members of the public, thus both parties will gain from improved efficiency and from efforts to address these concerns so as to enable timely investments.

Public input can also affect an extremely important factor of a proposed power plant project: its location, which, in turn, is a key determinant of project economics. Power plant promoters choose to locate new facilities at sites that are judged to be at or near optimum conditions from economic and operational points of view. Increasingly, they also carefully take into account environmental and other considerations. For example, a particular site may be selected because it provides an opportunity to minimise the investment requirement, is close to transmission lines, and can benefit from proper system voltage support. Any one of these criteria has significant influence on the others. For example, without adequate voltage support, the ability of the system to transfer energy would be reduced and the ability to supply energy to loads would be lessened. Thus, capability of the power system can only be maintained or improved by adding generating capacity in the right amounts, and in the most effective locations. In some cases, public opposition has forced regulators to demand that investors find an alternate site.

Public interest is – and should be – a key guiding principle for regulatory actions. A key challenge for regulators is to assess whether a proposed project, at a given site, is in the public interest. Public interest includes all stakeholder interests, and implies efforts to achieve a balance of economic, environmental and social interests. Regulators are accountable for assessing the public good a project may create, and for evaluating its potential positive and negative aspects. A specific position or point of view put forth by an individual or local interest group must be weighted against the broader public interest. If the project is deemed to be in the broader public interest, the project should be given approval to go ahead and measures to mitigate local concerns should be developed.

There are no firm criteria for determining the public interest that will hold well in every situation. Regulators must ultimately reach a conclusion that is considered appropriate at a particular point in time, under given circumstances. In practical terms, regulators must assess many factors including the project's potential economic, social and other benefits, and then determine whether these balance or outweigh the costs and impacts on the environment, public health and safety, and other related matters.

In the context of plant licensing, the regulatory mandate may be interpreted as having a dual role: *i)* To protect parties that may be affected by the construction and operation of the proposed generation facilities; and *ii)* To enable the investments that are assessed to be in the overall public interest. In essence, the dual role implies that regulation must be pursued in a way that seeks to protect against the negative impacts of energy development while enabling desirable outcomes determined to be in the public interest. Over the years, regulators have gained considerable experience in the protect function, *i.e.* addressing concerns raised by parties on which energy projects will have impact. Increasingly, regulators need to focus more attention on developing the enabling aspect of the regulatory culture. This enabling function implies that once a project is deemed to be in the public interest, the regulator should actively facilitate construction within the approved terms and conditions. Often the conditions attached to approvals are obstacles in themselves for various reasons: because they are not clear; because compliance cannot be assessed; or because the conditions are written in a way that makes them unattainable (NEB, 2005-2006).

In the United States, some wind energy promoters have voiced serious concern about various state regulatory approval processes that allow almost anyone to intervene in a regulatory application. By opening the process to such a degree, regulators may create situations in which a vocal minority essentially "game" the system and systematically oppose wind energy projects.

Italy provides an example of a centralised approach to project approval. In 2002, the government introduced a mechanism called *Sblocal Centrali*, which aims to streamline the approval process for certain types of projects, including construction of new power plants and modification or re-powering of existing plants. In the past, such projects were subject to separate authorizations by various local authorities. The *Sblocal Centrali* now oversees a single authorization process under the responsibility of the Ministry of Productive Authorities. This simplified procedure reduces risks for projects and time delays, thereby giving investors added incentive to submit project proposals. Since its introduction, investors have filed more than 70 applications for a total of 67 700 MW of new capacity. As of May 2005, the *Sblocal Centrali* had already released 41 authorisations for a total of 35 GW. Twenty of these permits, totalling a capacity of 15 430 MW, were granted under the new legislative framework.

In 2004, Italy introduced the *Marzano Law*, designed to ensure timely approvals of energy infrastructure projects. This law obliges regional authorities to respect a maximum 180-day delay in replying to applications for authorisation of new energy infrastructure. If this delay cannot be respected, the law transfers the authority to the central government. While the effectiveness of this law remains to be seen, these initiatives are good practices to streamlining licensing procedures for energy infrastructures (IEA, 2005d).

The Irish government recently enacted legislation to address energy planning issues. *The Planning and Development (Strategic Infrastructure) Act 2006* provides for a streamlined planning process in respect to certain projects of strategic importance to the state. For example, the act provides for specific new planning procedures for electricity transmission lines and strategic gas developments (including pipelines and terminals). It also provides a list of strategic energy infrastructure projects (including generation stations, LNG terminals, etc.) that can benefit from the new streamlined process. The Irish government is also considering the establishment of a state "bank" of sites (*i.e.* sites reserved for energy projects) for infrastructure development. Denmark has a similar "banking" system for wind farm development.

It is in the best interest of regulators to involve stakeholders in earliest possible stages of project planning, particularly as a means of raising awareness of the project as well as identifying the key issues that need to be addressed. Early involvement creates opportunities for early resolution of disputes, if any conflict arises. The overall objective is to build local acceptance and reduce risks through improved "predictability" – *i.e.* to avoid surprises on the part of the project promoters and other stakeholders. Typically, the stakeholders want to ensure that the regulator is striving to achieve an appropriate balance between basic property rights, democratic rights and the need for infrastructure – and is making its decisions in the overall public interest.

Effective regulatory practices that support early-stage interaction include clearly defined procedures, opportunity for local involvement, and transparent and coherent allocation of responsibilities between the involved authorities. Alberta (Canada) provides an example of how an early-stage interaction approach can be implemented through the creation of "synergies groups". Essentially, this is a mechanism through which the public, landowners, industry, and the provincial regulator can come together to discuss energy development issues and develop local solutions. The goal is to use interest-based negotiation to resolve the community's issues and concerns regarding a particular energy project. The idea is for the regulator to bring both sides – the company and the community – together with the aim of reaching agreement on the terms under which a company is granted a "social license" to operate. This social license is not an actual license; rather it serves as a tacit agreement between the community and the company to continue to participate in an engaged, meaningful dialogue – usually across all key issues and for the entire lifespan of the project.

Another approach to address citizen concerns and gain wider public acceptance is to provide direct compensation to the individuals or communities that may be adversely affected by an energy project and that therefore may have an interest in opposing it. In the UK, the *Barker Review of Land Use Planning,* in its final report to government in December 2006, pointed out that the current infrastructure planning system offers relatively little in the way of direct financial benefits for authorities to enhance local well-being and prosperity and that government should consider reforms to the funding system which would enable local authorities to share in the benefits of economic growth (Barker, 2006).

Essentially, the contention is not that developers should compensate for any loss of welfare. It is rather that those developers who chose to offer incentives to individuals or communities affected in order to gain wider acceptance for their project should not face unnecessary restrictions in doing so. There will be some who will be unwilling to accept the project at any price. They can continue to respond to a public enquiry process setting out the reasons for their objections, and the hearings process should continue to ensure that the project applications are properly assessed.

Such compensation schemes operate already in a number of jurisdictions, sometimes in the form of "good will" or "direct community benefit" payments. For example, developers and operators of offshore wind farms in Scotland have established a good-will payment scheme for local residents affected by the development. In most cases, these have been administered by independent local trusts, responsible for the collection and disbursements of payments. In France, an alternative model operates, where the central government can offer direct payments to authorities to take unpopular infrastructure development. New Zealand allows for compensating side-payments to be made by developers to those who may be negatively affected by a proposed development.

Key Message

The regulatory system should facilitate public participation and inputs as early as possible in the approval process. Regulators need to exercise vigilance in assessing minority versus broader public interest. Regulatory efficiency can be enhanced by creating a central regulatory authority or a "one-stop shop" regulatory entity. Establishing a "bank" of sites destined for energy infrastructure development is highly recommended as a means of fast-tracking approval of generation investments. Direct compensation to those affected by projects can be justified on reasons of fairness while they may also enable faster and wider public acceptance of projects.

Incorporating stakeholders' interests in early stages of approval is an appropriate approach to resolve issues early in the process and thus can reduce regulatory delays. One way to acknowledge the importance of stakeholder involvement is to encourage a culture of open dialogue based on the idea that regulators grant permits while the communities grant permission. Several key elements are necessary to effectively address public concerns: a clear description of the regulatory process (from application to final decision); a system ensuring respect for basic property rights and mechanisms for fair compensation for devalued property; and clear definition of the roles and responsibilities of various authorities involved, including a mechanism to expedite the process in case of unresolved issues.

Facilitating the Nuclear Option

Renewed interest in nuclear power is evident in many IEA countries in recent years. It often reflects concerns regarding energy security, surging fossil-fuel prices and rising CO_2 emissions. Nuclear power is a proven technology for large base-load electricity generation. Its generating costs are less vulnerable to fuel price changes than coal- or gas-fired generation. In addition, uranium resources are abundant and widely distributed globally. IEA estimates indicate that the operation of 1 GW of nuclear power generating capacity as a replacement for coal-fired generation reduces CO_2 emissions by 5 to 6 million tonnes per year.

The *WEO 2006* projects that by 2030 OECD nuclear capacity will amount to between 296 GW (Reference Scenario) and 362 GW (Alternative Scenario), compared to 308 GW in 2005. High capital costs - USD 2 to 3 billion per reactor unit – are a key challenge for nuclear power plant projects, making it significantly more difficult to acquire financing than for other technologies.

Long lead times in the planning and licensing phase, as well as in the construction phase, present another key risk factor. Countries with nuclear facilities already in place can expect a total lead time of 7 to 15 years, from investment decision to commercial operation. Countries with no experience in nuclear operation require even longer lead times. Construction times for nuclear power plants are substantially longer than those for CCGT plants (two to three years), wind power plants (one to two years) and coal-fired plants (four years). International experience reveals that nuclear power plants construction in several jurisdictions has encountered delays due to disputes and local opposition regarding plant licensing and siting, difficulties in securing access to water for cooling, and problems arising from other technical and management issues.

The United States provides an example where the federal government has adopted major initiatives to support the construction of new nuclear power plants. In 2002, the US government initiated the Nuclear Power 2010 Program with the aim of streamlining the regulatory process for the construction and operation of new nuclear power plants. In the past, for most of the existing nuclear power plants built between 1965 and 1985, the US Nuclear Regulatory Commission (NRC) issued a construction permit based on a preliminary design; it did not attempt to address or resolve safety issues until the plant was essentially complete. This former approach caused considerable delays for many nuclear projects.

The new nuclear licensing process improves the previous system by establishing three main phases: early site approval; design certification; and a combined license for construction and operation. Table 4.1 compares the licensing process under the old and new systems.

Table 4.1

New US nuclear licensing process improves previous system

Before	Now
Changing regulatory standards and requirements	More stable process: NRC approves site and design, single license to build and operate, before construction begins and significant capital is placed at risk
Design as you build	Plant fully designed before construction begins
No design standardisation	Standard NRC-certified designs
Inefficient construction practices	Lessons learned from nuclear construction projects overseas incorporated; modular construction practices
Multiple opportunities to intervene; cause delays	Opportunities to intervene limited to well-defined points in process; must be based on objective evidence that requirements have not been, and will not be, met

Source: Nuclear Energy Institute, 2007.

The Early Site Permit process allows an energy company to obtain federal regulatory approval for a new nuclear plant site before making a final decision to build a plant. The company can "bank" or reserve the site for up to 20 years. When the company is ready to build on a pre-approved site, it can proceed with choosing a power plant design and obtaining regulatory approval for construction and operation. The NRC is expected to grant up to 3 Early Site Permits in 2007 in addition to one already issued.

The Design Certification process allows plant designers to secure advance NRC approval of standard plant designs. A nuclear power company can then select a design from those that are already NRC-approved. This process makes it possible for the public to review and comment at the design stage – i.e. long before any construction begins. The NRC design certification process aims to fully resolve safety issues associated with the proposed design, which is then "approved" for 15 years. According to the US Energy Information Administration, as of November 2006, the NRC has certified the following reactor designs: the AP600 (650 MW), AP1000 (1117 MW) and System 80+ (1300 MW) by Westinghouse; and the ABWR (1371 MW) by a consortium led by General Electric. Several other designs are undergoing certification or pre-certification.

Once a company has selected a site and a design, it applies for a Combined License for Construction and Operation (COLA), at which time operational and site-specific design details are incorporated. All issues resolved in connection with earlier proceedings on siting and design are considered "closed" at this point – i.e. they cannot be brought forward again during the combined license proceeding. This firm delineation of issues and phases contributes to a more effective regulatory process. The Nuclear Energy Institute estimates that as of February 2007, there are 15 companies and consortia preparing license applications to build and operate as many as 33 new nuclear reactors, or approximately 43 GW of generating capacity. The first application for COLA will likely be filed by the fourth quarter 2007.

The new nuclear licensing approach in the US assumes that reactors will be built in "families" of the same design, with the exception of a limited number of site-specific differences. Such standardisation will likely reduce overall design, planning and construction costs, and lead to greater efficiencies and simplicity in nuclear plant operations, including safety, maintenance, training and spare-parts procurement. It may also result in "standardised available" technologies rather than "experimental" approaches to achieve the best technology. Real cross-border benefits might be realised if certain designs are deemed internationally acceptable, as opposed to having to demonstrate safety to regulators in multiple countries.

Box 4.2. US Energy Policy Act of 2005 and nuclear power

The US Energy Policy Act of 2005 *includes specific incentives to encourage investments in new nuclear plants:*

- *Federal loan guarantees – up to 80% of the project cost – for advanced nuclear reactors or other emissions-free technologies. This loan guarantee should allow companies building nuclear plants to apply a more leveraged capital structure than is typical of regulated utility financing.*

- *Production tax credits of USD 0.018/KWh for the first 6 000 MW of new nuclear capacity in their first eight years of operation. The credits will be distributed on a pro-rata basis to all plants that: a) submit a COLA application to the NRC by 31 December 2008, b) begin construction by 1 January, 2014 and c) start commercial operation by 1 January 2021.*

- *Federal insurance coverage for delays caused by licensing or litigation. This standby support covers the first six plants: first two plants have USD 500 million policies covering 100% share of costs resulting from delays, and second four plants have USD 250 million policies covering 50% of delay costs.*

- *A 20-year extension of the* Price-Anderson Act *for nuclear liability protection.*

- *Support for advanced nuclear technology.*

Delays in the regulatory process represent a risk that is beyond industry control. Since the new licensing process has not yet been fully applied and tested, potential remains for encountering costly pitfalls in the approval of new nuclear plant projects. To alleviate the risks for regulatory delays, the *Energy Policy Act of 2005* (signed into law in August 2005) provides some form of investment protection. Under this Act, the federal government will indemnify debt service and other costs for the first six new plants if commercial operation is delayed for reasons beyond the company's control, such as litigation or a failure by the NRC to meet license review schedules (Box 4.2).

The United Kingdom is in the process of overhauling rules related to planning inquiries for large-scale electricity projects. The reforms, which are set out in the Energy Review, will streamline the planning system and aim at greater efficiency and greater transparency. Under the current approach, nuclear site licenses are granted by the Health and Safety Executive (HSE); the HSE's Nuclear Installations Inspectorate (NII) administers the licensing function. The NII will not grant a nuclear site license unless it is satisfied that a prospective operator has the capacity to meet all stringent safety requirements – from design through to decommissioning – in adherence to conditions attached to the site license. Discharges of radioactive material (in gaseous or liquid form) to the environment from licensed sites are strictly controlled, as is the disposal of solid radioactive wastes. In England and Wales, discharge authorizations are administered by the Environment Agency; in Scotland, by the Scottish Environment Protection Agency. The NII collaborates closely with the Environment Agency and the Scottish Environment Protection Agency under the terms of a Memorandum of Understanding that sets out the lead roles of the organisations, as well as requirements for liaison and consultation.

The UK is also developing a new policy and regulatory framework for new nuclear construction. In July 2006, the Department of Trade and Industry issued a consultation document titled *Policy Framework for New Nuclear Build* (DTI, 2006). The document acknowledges that past procedures led to an inefficient system and created cost and uncertainty for all system participants. The primary problem was that scope was too broad in that planning included discussions on strategic national and regulatory issues as well as project specific and local issues. Key elements of the proposed framework include:

- A "Statement of Need" for nuclear power will be incorporated in the national government policy.

- A justification process, to be established under the responsibility of the Secretary of State for Trade and Industry. The government is proposing that planning inquiries should not consider generic questions related to the health and safety aspects of nuclear power (*e.g. Is nuclear power safe?*). Rather, planning inquiries should proceed on the assumption that the relevant evidence on these topics has been considered as part of the justification decision by the Secretary of State.

- A government-led strategic assessment, involving public consultation, to determine the high-level environmental impacts of new nuclear facilities. Fully engaging the public at this strategic level will eliminate the need to re-assess the same large-scale considerations at later public inquiries that are site specific.

- A planning inquiry that focuses on the current proposal and its relationship to local plans and to local environmental impacts. This inquiry would involve public input and would take into account the other national or strategic considerations.

- New inquiry rules will be introduced to support the new policy and regulatory framework.

The new policy framework on new nuclear development is expected to be included in a forthcoming Energy White Paper. The proposed generic design and site licensing processes are presented in Figures 4.1 and 4.2.

Figure 4.3 maps the major steps in licensing a new nuclear power plant in Canada. Separate licences are required for each of the five phases in the plant life cycle: *i)* site preparation; *ii)* construction; *iii)* operation; *iv)* decommission; and *v)* abandoning. Under the *Nuclear Safety and Control Act (NSCA)*, the Canadian Nuclear Safety Commission (CNSC) is responsible for regulating all nuclear facilities and nuclear-related activities. The CNSC serves as both an independent federal regulatory agency and a quasi-judicial administrative tribunal. Prior to issuing a license, the Commission requires the company to complete a comprehensive environmental assessment pursuant to the *Canadian Environmental Assessment Act*, which includes mandatory opportunities for public participation. The Commission may not issue a licence unless it is satisfied that the applicant will make adequate provisions to protect health, safety, security and the environment, and to implement international obligations to which Canada has agreed.

In France, nuclear law is spread across many pieces of legislation. In 1963, the government introduced a system for licensing and controlling major nuclear installations that gave the government itself responsibility in matters of population and occupational safety (Decree of 11 December 1963). The Minister for Industry and the Minister for Ecology and Sustainable Development are the main authorities involved in the licensing for large nuclear installations. The licensing procedure is governed by Decree No. 63-1128 of 11 December 1963. Under this procedure, the decree authorizing the installation sets out the technical requirements and other formalities with which the plant operator must comply. For nuclear reactors, there are generally two stages: first, fuel loading and commissioning tests; and second, entry into operation. Both stages are conditional on joint approval by the Minister for Industry and the Minister for Ecology and Sustainable Development. The consent of the Minister for Health is also requested. The French nuclear programme provides an example of the gains achieved through standard design. Over nearly two decades, France built

Figure 4.1

Generic design assessments: Regulatory processes for new nuclear power stations in the United Kingdom

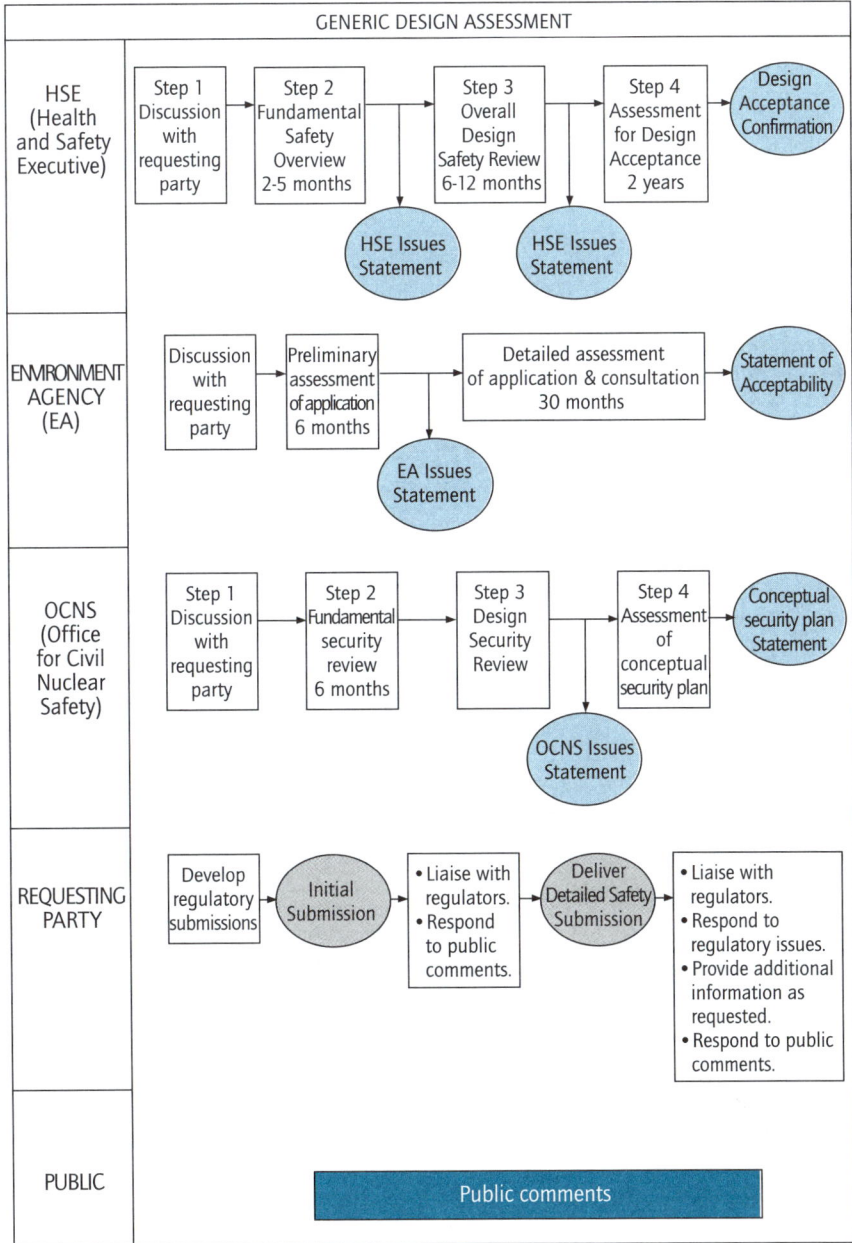

OUTLINE TIMETABLE

Source: HSE, 2007.

Figure 4.2

Figure 4.2 Nuclear site licensing: Regulatory processes in the United Kingdom

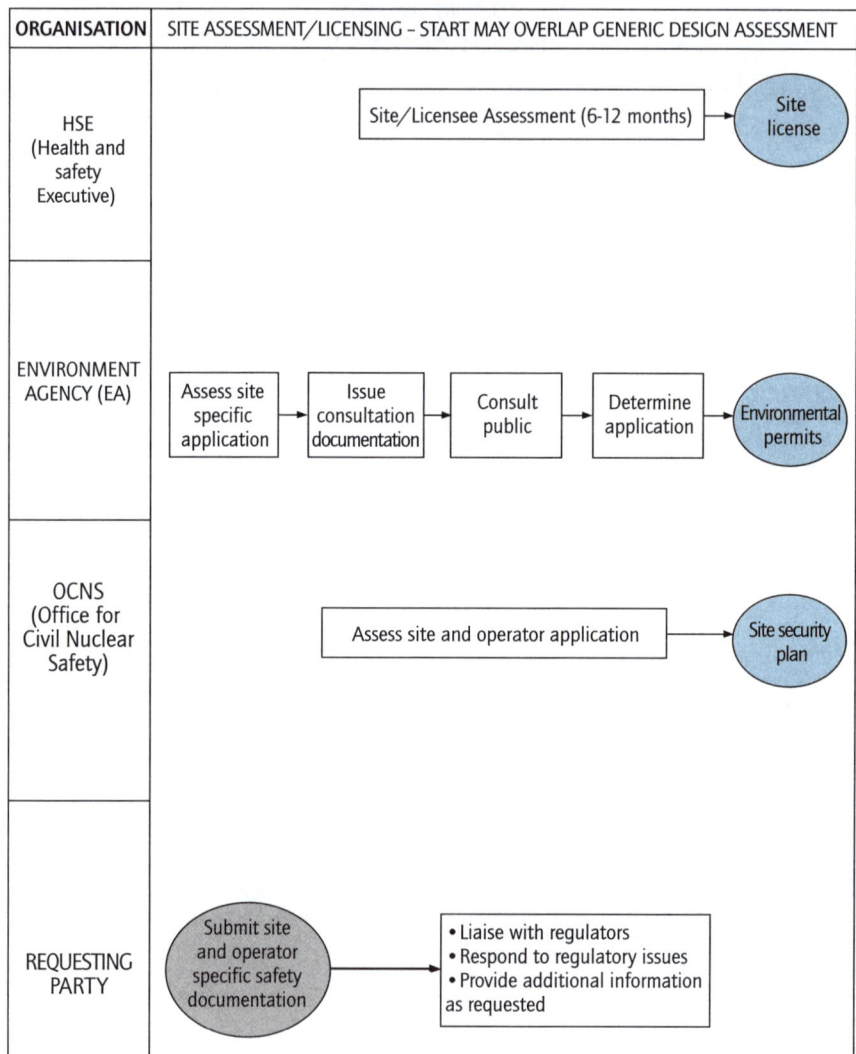

Source: HSE, 2007.

Figure 4.3

Licensing of new nuclear power plants in Canada

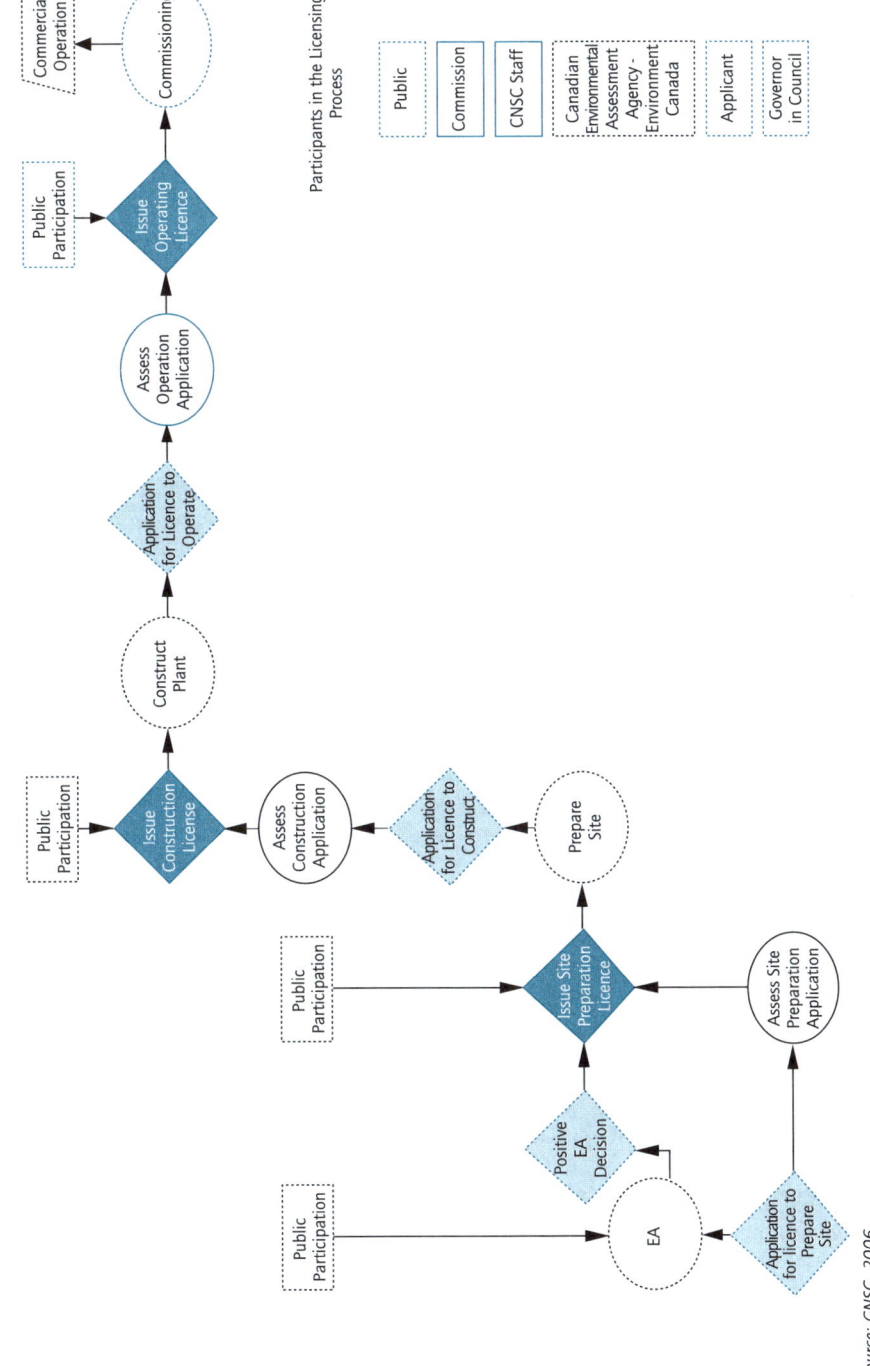

Source: CNSC, 2006.

34 standardised 900 MW reactors and 20 reactors of 1 300 MW capacity. The first reactor took about seven years to build; the last reactor took only five years.

In Korea, the licensing procedures for new nuclear power plants are divided into three stages: site selection; construction permit; and operating license. In the site selection stage, the conceptual design is examined to determine the appropriateness of the proposed site. At this stage, the safety requirements of the site are reviewed from the standpoints of the design, construction, and operation of the plant. To obtain the construction permit, the utility must submit a Preliminary Safety Analysis Report (PSAR) and an overall quality assurance programme for the project along with the reference design of the plant. This stage also requires that the utility/project promoter prepare an environmental impact statement. When the utility requests an operating license, the Ministry of Science and Technology (MOST) must confirm that the as-built plant conforms to the submitted design. In this stage, the company submits operational and technical specifications as well as emergency plans and procedures against radiation hazards. MOST has the overall responsibility for ensuring the protection of public health and safety through regulatory control and safety inspections, based on the provisions of the *Atomic Energy Act*. Regulatory inspections of nuclear power plants under construction or in operation are implemented in five ways: *i)* according to the procedure of a pre-operational inspection of the nuclear installation; *ii)* through periodic inspections of the operating nuclear installations; *iii)* via quality assurance audits; *iv)* in the context of daily inspections by resident inspectors; and *v)* under special inspections.

In Japan, the licensing process for nuclear power plants comprises three main steps (Figure 4.4): application for site selection, application for reactor installation, and application to construct. Local public opinion can be voiced in public hearings that are held during the first two phases. The site selection requires an environmental assessment. A permit for nuclear reactor may be granted by the Atomic Energy Commission with input from the Nuclear Safety Commission and the consent of the Ministry of Education, Culture, Sports, Science and Technology.

Enabling efficient investments requires, among other things, creating an environment that makes it easier for potential licensees to get their application submission "right the first time". To this end, regulatory authorities must ensure that published regulatory requirements and standards are clear, that procedures and stakeholder intervention rules are well understood, and that regulatory decision making is consistent and predictable. This is crucially

Figure 4.4

Approval process for nuclear power plants in Japan

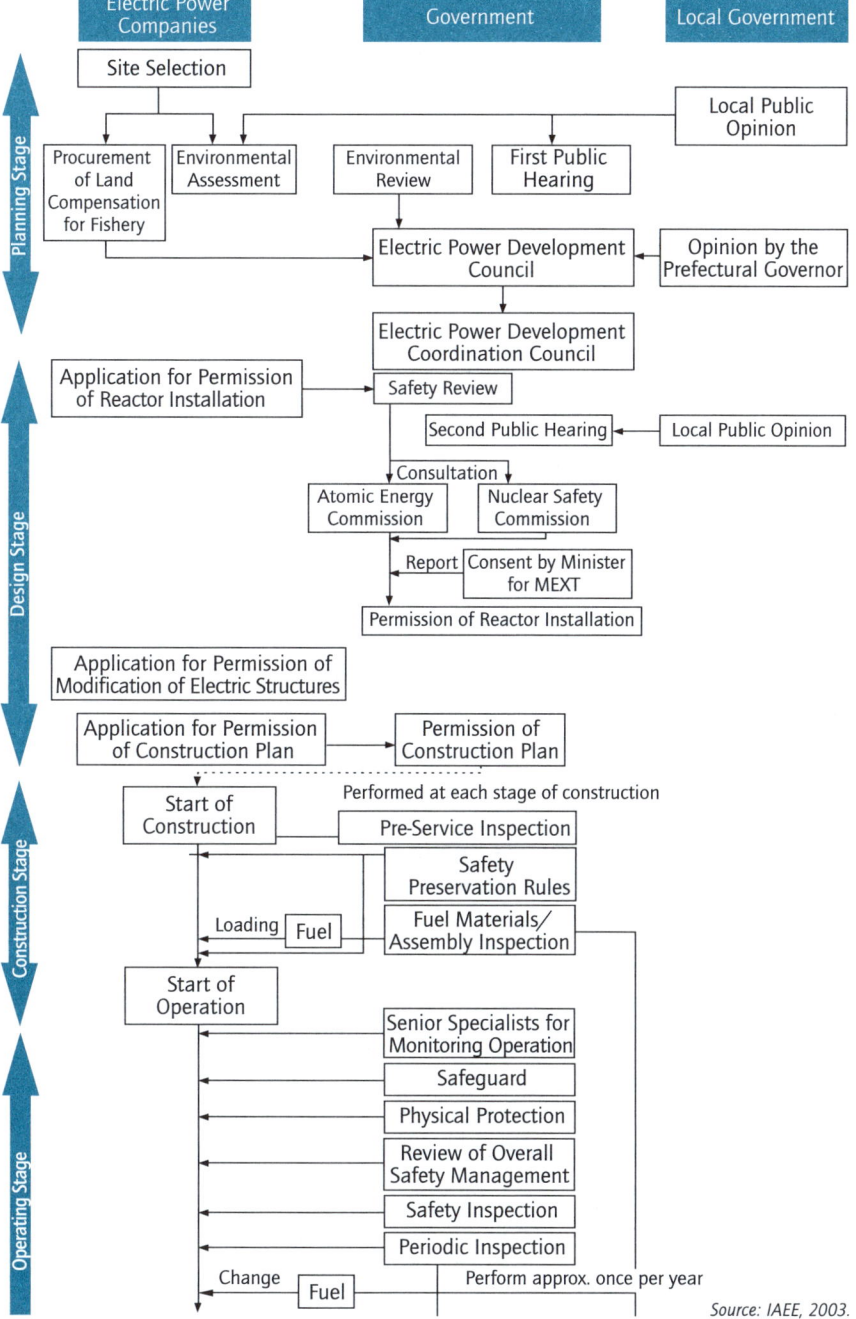

Source: IAEE, 2003.

important, particularly in that some nuclear licensing processes (*e.g.* in the United States) have not demonstrated adequate support for business decisions to build new nuclear power plants. Governments and regulators must ensure that all parties understand the entire process for new nuclear plant licensing applications. This includes ensuring that the technical content and assessment criteria are properly developed and documented. In view of the projected increase in applications in several jurisdictions, nuclear regulatory agencies should be proactive in terms of staffing and planning to ensure adequate capability to manage the expected increase in workload.

Management and disposal of nuclear waste remains one of the most contentious issues for regulators and other stakeholders. Thus, a key aspect of regulating nuclear power involves the development of a clear, stable and predictable framework for radioactive waste storage and long-term disposal. Currently, most spent nuclear fuels and radioactive by-products are stored at reactor sites. There seems to be a scientific consensus that deep geologic repository offers the best solution for long-term deposition. Comprehensive studies carried out by the US Department of Energy confirmed the feasibility of a national, geologic repository for used nuclear fuel at Yucca Mountain, Nevada. This site will need a license to build and operate from the Nuclear Regulatory Commission, and will take ten years or more to build. At present, its capacity is subject to a statutory limit of 70 000 metric tons of nuclear waste materials.

Having an "approved" central repository can help to reduce uncertainties for new investments in nuclear power plants: It suggests that the generic issue of permanent nuclear waste will have been addressed and resolved, partially or totally. In conjunction with the national repository, the development of a viable used fuel management strategy would help to improve perceptions of risk and increase public confidence in nuclear power plant operations. Without a clear solution regarding permanent storage, investors may perceive that regulatory risks are too high and be unwilling to commit.

Several other IEA countries are pursuing efforts to develop geological repositories. Sweden is currently in the process of selecting an appropriate site; candidate sites are close to two of the country's four nuclear power plants. Investors are likely to submit an application for repository construction in 2008, with the aim of getting the facility operational by 2017.

France, which uses about 12 400 tonnes of uranium per year for nuclear generation, also has an R&D programme covering deep geological disposal (including lab tests) and extended interim storage. At present, the country's spent fuels are reprocessed to recover useful fuel components and reduce the volume of high-level waste. Reprocessing is done at the La Hague plant (Normandie), where nuclear waste is stored for later disposal.

Finland is already constructing a deep rock laboratory to demonstrate the acceptability of the location for future construction of a deep repository. It currently operates a surface pool storage facility for spent fuel at the Olkiluoto nuclear power plant. The facility opened in 1987, has a capacity of 1 270 tonnes and is designed to hold spent fuel for about 50 years.

The United Kingdom has full fuel-cycle facilities, including major reprocessing plants. Most low-level waste is currently disposed of at the state-owned repository near Drigg in Cumbria. Intermediate level waste is stored mostly at the sites of nuclear facilities in operation, and eventually will be disposed of in a dedicated repository. High-level waste is stored at Sellafield, some vitrified and stored in stainless steel canisters in silos. The treatment of long-term waste management was more directly addressed in a July 2006 report *Managing our Radioactive Waste Safely*, which recommends deep underground, geological disposal of radioactive waste.

Key Message

Regulation and nuclear policies need to effectively address the key issues of safety, radioactive waste management and disposal, and decommissioning and related costs. Given the complexity of nuclear licensing, governments should aim to simplify and streamline the licensing process. Plant design standardisation and a multi-stage approach to plant licensing, allowing for early public inputs, can help to improve regulatory efficiency by reducing processing time, costs and uncertainty.

It would be appropriate for regulators to commit to a reasonable, pre-established schedule for review and approval. This will improve stakeholder confidence. Further international co-operation should be explored with the aim of establishing an internationally acceptable knowledge with respect to technology and standards, as well as the safety and health impacts of nuclear operations and waste disposal and management. This would be a highly desirable means of avoiding costly and time-consuming duplication of regulatory efforts.

REFERENCES

Agersbaek, Gitte, 2006, "Renewable Energy: How much can we expect to increase supplies over the next two decades?", presentation at the International Council of Capital Formation workshop on EU energy supply, October 2005, www.iccfglobal.org

Auer, Hans, Michael Stadler, Gustav Resh, Claus Huber, Thomas Schuster, Hans Taus, Lars Henrik Nielsen, John Tidwell and Derk Jan Swider, 2004, "Cost and Technical Considerations of RES-E Grid Integration", a report in the GreenNet study, www.greennet.at

Barker, Kate, 2006, "Barker Review of Land Use Planning: Final Report – Recommendations", www.barkerreviewofplanning.org.uk

Blyth, W. and M. Yang, 2006, "Impact of climate change policy uncertainty in power investment", IEA Working Paper Series LTO/2006/02

Bushnell, James, 2005, "Electricity Resource Adequacy: Matching Policies and Goals", *CSEM Working Papers*, vol. 146, Center for the Study of Energy Markets at University of California Energy Institute, www.ucei.org

Caramanis, M.C., 1982, "Investment decisions and long-term planning under electricity spot pricing", IEEE, *Transactions on Power Apparatus and Systems*, vol. 101 (12), pp. 4640-4648

Canadian Nuclear Safety Commission (CNSC), 2006, "Licensing Process for New Nuclear Power Plants in Canada", Infor-0756, www.nuclearsafety.gc.ca

Cramton, Peter and Steven Stoft, 2006, "The Convergence of Market Design for Adequate Generation Capacity with Special Attention to the CAISO's Resource Adequacy Problem", *CEEPR Working Papers*, Vol. 2006-07, MIT Center for Energy and Environmental Policy Research, web.mit.edu/ceepr/www

De Joode, Jeroen, Douwe Kingma, Mark Lijesen, Machiel Mulder and Victoria Shestalova, 2004, "Energy Policies and Risks on Energy Markets: A cost-benefit analysis", CPB Netherlands Bureau for Economic Policy Analysis, www.cpb.nl

Department of Trade and Industry (DTI), 2006, "Policy Framework for New Nuclear Build – Consultation Paper", www.dti.gov.uk

Dixit, A. K. and R.S. Pindyck, 1994, *"Investment under Uncertainty"*, Princeton University Press

Edleson, M. and F., Reinhart, 1995, "Investment in pollution compliance options: the case of Georgia Power", *Real Options in Capital Investment: Models, Strategies and Applications,* edited by L. Trigeorgis Praeger Publishers, Westport

EPRI, 1999, *"A Framework for Hedging the Risk of Greenhouse Gas Regulations"*, EPRI, Palo Alto

ERGEG, 2006, "European Regulators' Interim Report on the 4th November Black-out in Europe", www.ergeg.org

European Commission, 2005, "Commissioner Pibalgs: Europe could save 20% of its energy by 2020", EC press release, Brussels, 22 June 2005, www.ec.europa.eu/energy

European Commission, 2006, *"Energy Sector Inquiry; Draft Preliminary Report"*, EC DG Competition, Brussels

EWEA (European Wind Energy Association), 2006, "Wind Power installed in Europe by the end of 2006 (cumulative)", statistics on www.ewea.org

ERGEG, 2006a, "ERGEG Guidelines for Good Practice on Information Management and Transparency in Electricity Markets", Ref. E05-EMK-06-10, www.ergeg.org

ERGEG, 2006b, "ERGEG interim report on the lessons to be learned from the large disturbance in European power supply on 4 november 2006", Ref. E06-BAG-01-05, www.ergeg.org

ETSO, 2006, "Generation Adequacy: An Assessment of the Interconnected European Power Systems 2008-2015", www.etso-net.org

Evans, Joanne and Richard Green, 2005, "Why did British electricity prices fall after 1998?", *CEEPR Working Papers,* Vol. 2003-07 (revised 2005), MIT Center for Energy and Environmental Policy Research, web.mit.edu/ceepr/www

FERC, 2006, "Assessment of Demand Response & Advanced Metering", staff report, docket number AD-06-2-000, www.ferc.gov

Folketinget, 2003, "Besvarelse af spørgsmål 41-45 stillet af Det Energipolitiske Udvalg den 19. December 2002", translated: "Response to questions 41-45 from the Energy Policy Commission on 19 December 2002", www.folketinget.dk

Frayer J. and N.Z. Uludere, 2001, "What is it worth? Application of real options theory to the valuation of generation assets", *Electricity Journal,* 14 (8) p40-51, Elsevier Inc.

Health and Safety Executive (HSE), 2007, "Guide to Regulatory Processes for Generic Design Assessment of New Nuclear Station Build", www.hse.gov.uk

Hurlbut, David, Keith Rogas and Shmuel Oren, 2004, "Protecting the Market from "Hockey Stick" Pricing: How the Public utility Commission of Texas is Dealing with potential Price Gouging", *The Electricity Journal,* vol. 17.3, pp. 26-33, Elsevier Inc.

IAEA, 2003, "Country Nuclear Power Profiles", www.iaea.org

IEA, 2002, *Security of Supply in Electricity Markets,* OECD/IEA, Paris

IEA, 2003, *Power Generation Investment in Electricity Markets,* OECD/IEA, Paris

IEA, 2004a, *World Energy Outlook,* OECD/IEA, Paris

IEA, 2004b, *Prospects for CO2 Capture and Storage,* OECD/IEA, Paris

IEA, 2005a, *Lessons From Liberalised Electricity Markets,* OECD/IEA, Paris

IEA, 2005b, *Learning From the Blackouts: Transmission System Security in Competitive Markets,* OECD/IEA, Paris

IEA, 2005c, *Energy Policies of IEA Countries: Spain,* OECD/IEA, Paris

IEA, 2005d, *Energy Policies of IEA Countries: 2005 Review,* OECD/IEA, Paris

IEA, 2006a, *Energy Policies of IEA Countries: New Zealand,* OECD/IEA, Paris

IEA, 2006b, *World Energy Outlook 2006,* OECD/IEA, Paris

IEA, 2006c, *Energy Policy Perspectives: Scenarios and Strategies to 2050,* OECD/IEA, Paris

IEA, 2006d, *Natural Gas Market Review 2006: Towards a Global Gas Market*, OECD/IEA, Paris

IEA, 2006e, *Energy Policies of IEA Countries: the United Kingdom*, OECD/IEA, Paris

IEA, 2006f, *Electricity Information*, OECD/IEA, Paris

IEA, 2007, *Climate Policy Uncertainty and Investment Risk*, OECD/IEA, Paris, forthcoming

Ishii, J. and J. Yan, 2004, "Investment under regulatory uncertainty: US electricity generation investment since 1996, *CSEM Working Paper*, 127, Center for the Study of Energy Markets, University of California Energy Institute, www.ucei.org.

Joskow, Paul, 2006, Competitive Electricity Markets and Investment in New Generating Capacity, in *The New Energy Paradigm*, Oxford University Press

Lambrecht, B. and W. Perraudin, 2003, "Real options and pre-emption under incomplete information", *Journal of Economic Dynamics and Control*, Elsevier

Laurikka, H and T. Koljonen, 2006, "Emissions trading and investment decisions in the power sector – a case study in Finland", *Energy Policy*, 34(9) p1063-1074, Elsevier

National Energy Board, 2005-2006, Speeches and presentations, www.neb-one.gc.ca

National Grid, 2004, "2004 Seven Year Statement", www.nationalgrid.com

National Grid, 2006, "Revised Derivation of the Main Energy Imbalance Price", Balance and Settlement Code Modification Proposal P194, www.ofgem.gov.uk

NEMMCO, 2006, "2006 Statement of Opportunity", www.nemmco.com.au

NEA, 2003, *"Decommissioning Nuclear Power Plants"*, NEA/OECD, Paris

NEA, 2006, *"Forty Years of Uranium Resources, Production and Demand in Perspective"*, OECD/NEA, Paris

NEA, 2007, "The Changing Climate for Nuclear Energy – Annual Briefing for the Financial Community", www.nea.fr

NEA/IAEA, 2005, *"Uranium 2005: Resources, Production and Demand"*, OECD/NEA and IAEA, Paris

NEA/IEA, 2005, *"Projected Costs of Generating Electricity – 2005 Update"*, OECD(NEA)/IEA, Paris

New York Times, 2006, "In Deregulation, Power Plants Turn into Blue Chips", 23 October 2006

Nordic Competition Authorities, 2003, "A Powerful Competition Policy: Towards a More Coherent Competition Policy in the Nordic Market for Electric Power", www.ks.dk

North west Power and Conservation Council, 2006, "A Pilot Capacity Adequacy Standard fot the Pacific North west", www.nwcouncil.org

Oren, Shmuel, 2005, "Generation Adequacy via Call Options Obligations: Safe Passage to the Promised Land", *Electricity Journal*, vol. 18, Elsevier Inc.

OXERA, 2005, "Margin for Error? Security of Supply in Electricity", *Agenda*, November 2005, www.oxera.com

PJM, 2006, "2005 State of the Market Report", www.pjm.com

Platts, 2006, "Bulgaria opts for Russian vendor to build two VVERs at Belene", *Nucleonics Week*, vol. 47, no. 44, McGraw-Hill

Reedman, L., P. Graham and P. Coombes, 2006, "Using a real-options approach to model technology adoption under carbon price uncertainty: an application to the Australian electricity generation sector", *The Economic Record*, 82 Special issue, pp. 64-73, Blackwell Publishing

Robson, Nigel, 2006, "Quantifying Construction Risk", presentation at Euromoney conference Nucelar Energy Finance Forum, 31 October 2006, London

Roques, Fabien, William J. Nutall and David M. Newbery, 2006, «Using Probabilistic Analysis to Value Power Generation Investments under Uncertainty», *Cambridge Working Papers in Economics*, no. 0650, www.econ.cam.ac.uk

Rothwell, G., 2006, "A real options approach to evaluating new nuclear power plants", *The Energy Journal*, Vol. 27 No. 1 p37, IAEE

Schubert, Eric S., David Hurlbut, Parviz Adib and Shmuel Oren, 2006, "The Texas Eenrgy-Only Resource Adequacy Mechanism", *The Electricity Journal*, vol. 19.10, pp. 39-49, Elsevier Inc.

Sekar, R. et. al., 2005, "Future carbon regulations and current investments in alternative coal-fired power plant designs", MIT Joint Program on the Science and Policy of Global Change, report no. 129, web.mit.edu.

Svenska Kraftnät, 2002, "Effektförsörjningen på den öppna elmarknaden", (translated: "Capacity supply in the open electricity market"), report to the Swedish government, www.svk.se

Svenska Kraftnät, 2005, "Den svenska effektbalancen vintrarna 2004/2005 och 2005/2006" (translated: "The Swedish power balance during the winters 2004/2005 and 2005/2006"), report to the Swedish Ministry of Sustainable Development, www.svk.se

Svenska Kraftnät, 2006, "Effekttilgänglighet efter februari 2008", (translation: "Capacity availability after February 2008"), report, www.svk.se

The Danish Competition Authority, 2005, "Elsam A/S' misbrug af dominerende stilling i form af høje elpriser" (translated: "Elsam A/S abuse of dominating position through high electricity prices"), Meeting of The Danish Competition Council 30 November 2005, www.ks.dk

Thon, Scott, 2005, "Alberta Electricity Industry Restructuring: Implications for Reliability", http://www.energetics.com/meetings/reliability/pdfs/thon.pdf

Trigeorgis, Lenos, 1996, *Real Options*, MIT Press

UCTE, 2006, "Interim Report: System Disturbance 4th November", www.ucte.org

VDEW, 2006, "Stromwirtschaft investiert in Versorgungssicherheit," press release october 2006, www.vdew.de

Walhain, Seb, 2006, "Capturing the Low Carbon Value of a Nuclear Investment", Conference paper at Euromoney Nuclear Energy Finance Forum 2006, www.euromoneyenergy.com

Wilson, John D. and Brian H. Potts, 2007, "The end of new source review?", *Public Utilities Fortnightly*, vol. 145, no. 2, pp. 50-55, Public Utilities Reports Inc.

The Online Bookshop

IEA PUBLICATIONS, 9, rue de la Fédération, 75739 Paris Cedex 15

PRINTED IN FRANCE BY ACTIS

(61 2007 13 1P1) ISBN 13 : 978-92-64-03007-7 - 2007